MXenes: Fundamentals and Applications

Edited by

Inamuddin[1,2,3], Rajender Boddula[4] and Abdullah M. Asiri[1,2]

[1]Centre of Excellence for Advanced Materials Research, King Abdulaziz University, Jeddah 21589, Saudi Arabia

[2]Chemistry Department, Faculty of Science, King Abdulaziz University, Jeddah 21589, Saudi Arabia

[3]Department of Applied Chemistry, Faculty of Engineering and Technology, Aligarh Muslim University, Aligarh-202 002, India

[4]CAS Key Laboratory of Nanosystem and Hierarchical Fabrication, National Center for Nanoscience and Technology, Beijing 100190, PR China

Copyright © 2019 by the authors

Published by **Materials Research Forum LLC**
Millersville, PA 17551, USA

Published as part of the book series
Materials Research Foundations
Volume 51 (2019)
ISSN 2471-8890 (Print)
ISSN 2471-8904 (Online)

Print ISBN 978-1-64490-024-6
eBook ISBN 978-1-64490-025-3

Distributed worldwide by

Materials Research Forum LLC
105 Springdale Lane
Millersville, PA 17551
USA
http://www.mrforum.com

Manufactured in the United States of America
10 9 8 7 6 5 4 3 2 1

Table of Contents

Preface

MXenes are known as 2D metal carbides, and nitrides and have attracted great attention in recent years due to their high electrical conductivity, good hydrophilicity, chemical stability, and ultrathin 2D sheet-like morphology. MXenes have recognized huge consideration in this decade. In the drive toward green and economic science, investigating proficient and stable 2D metal carbides and nitrides practically equivalent to graphene and transition metal dichalcogenides (TMDs) has been the hot topic of scientific research. MXenes have attracted attention for many applications, e.g., catalysis, membrane separation, supercapacitors, hybrid-ion capacitors, batteries, hydrogen storage, nanoelectronics, and sensors.

MXenes: Fundamentals and Applications introduce readers to fundamental aspects, properties, fabrication methods, and their numerous applications, with a strong focus on novel findings and technological challenges. This book provides an in-depth look of 2D metal carbides and nitrides and their key applications for electronics, catalysis, sensors, biomedical, energy, hydrogen storage, and environmental fields. The book is the result of commitments by top researchers in the field of MXenes with various backgrounds and expertises. This book is a one-stop reference for MXene materials and overviews up-to-date literature in the field of advanced Mxene materials and applications.

This is the first book on MXene materials and application in the market. It serves as invaluable guide to students, professors, scientists and research and development industrial specialists working in the field of advanced science, nanodevices, flexible electronics, energy, and environmental science.

Inamuddin[1,2,3], Rajender Boddula[4] and Abdullah M. Asiri[1,2]

[1]Chemistry Department, Faculty of Science, King Abdulaziz University, Jeddah 21589, Saudi Arabia.

[2]Centre of Excellence for Advanced Materials Research, King Abdulaziz University, Jeddah 21589, Saudi Arabia

[3]Department of Applied Chemistry, Faculty of Engineering and Technology, Aligarh Muslim University, Aligarh-202 002, India

[4]CAS Key Laboratory of Nanosystem and Hierarchical Fabrication, National Center for Nanoscience and Technology, Beijing 100190, PR China

MXenes: Fundamentals and Applications
Materials Research Foundations **51** (2019) 1-19

Materials Research Forum LLC
doi: https://doi.org/10.21741/9781644900253-1

Chapter 1

MXenes for Sensors

Matheus Costa Cichero[1], João Henrique Zimnoch Dos Santos*[1]

[1]Instituto de Química - Universidade Federal do Rio Grande do Sul, Av. Bento Gonçaves 9500, CEP 91501-970, Porto Alegre, RS, Brasil

*jhzds@iq.ufrgs.br

Abstract

MXenes are two-dimension materials based on transition metal carbides, nitrides or carbonitrides, which were first obtained from their precursor 3D bulk layered materials Ti_3AlC_2 MAX phase in 2011, resulting in Ti_3C_2. MXenes possess the metallic conductivity and hydrophilicity through the hydroxyl and oxygen surface terminations that results from the etching process, which in turn afford their use in different applications such as supercapacitors, batteries, microwave absorption, optical communication, catalysis and more. In this chapter some fundamentals are considered such as most employed synthetic route, as well as some of the post treatment commonly used. Recent examples of electronic and bioanalytical sensors are depicted, as well as instrumental techniques usually employed in their characterization.

Keywords

2D materials, MXenes, Sensors, Biosensors, Titanium Carbide

Contents

1.   Introduction

Two-dimension (2D) materials may be defined as those compounds that exhibit one to several atomic thick layers, which interact very weakly. Restricting one dimension to the nanometer scale, the resulting 2D materials exhibit different properties from their 3D counterpart, particularly in terms of mechanical and electronic characteristics like high surface-to-volume ratio, large band gaps, ultrathin thickness, and flexibility. Such unique mechanical, optical and electronic properties are very desirable in several application fields as energy, electronics, sensors devices just to mention a few [1–4]. Examples of 2D materials comprise: Hexagonal boron nitride (h-BN) [5,6], transition metal dichalcogenides (TMDs) [7–9], covalent-organic frameworks (COFs) [10,11], black phosphorus (BP) [12,13] among others, and probably the most popular and studied one is graphene [3,14–18]. A new class of 2D materials has been introduced – MXenes, which refers to a "graphene-like morphology" are materials that are composed of layers of transitions metal carbides, nitrides or carbonitrides [19]. MXenes applications comprise their use in the development of materials such as batteries, microwave absorption devices, catalysts for instance. An overview of some recent research applications is depicted in Table 1 [20-29].

Since the debut in 2011, many reviews devoted to MXenes have discussed their characteristics and applications. A broader panorama of recent potential uses of MXenes can be found in the review done by Li et al. [30] in which details concerning energy store field are provided: MXenes applied in lithium-ion battery electrodes and capacitors, non-lithium ion battery electrodes, supercapacitors and hydrogen storage are some examples of materials which are deeply discussed. The optical and electronic properties that are associated with the majority of MXenes applications are highlighted in the text of Hantanasirisakul et al. [31] both within the theoretical and experimental approach. Their study serves as a guide of the current state-of-art of MXenes structure and properties. The review by Anasori et al. [32] describes in details the synthesis process, comprising etching and delamination processes as well the different structures and proprieties. The review by Zhu et al. [33] also focuses on synthesis issues, including etching and delamination processes aiming at obtaining single and multilayer MXenes. Their discussions also comprise stability, mechanical, electronic and surface chemical properties of MXenes up to their uses as a catalyst for several reactions. It is worth here mentioning that one of the properties of MXenes concern their capacity to easily

MXenes: Fundamentals and Applications   Materials Research Forum LLC
Materials Research Foundations **51** (2019) 1-19   doi: https://doi.org/10.21741/9781644900253-1

incorporate other molecules due to their hydrophilic and biocompatible characteristics. Such behaviors allow their use as a catalyst support or agents delivering systems, for instance. The employment of MXenes in composites is subjected to the review by Sinha and co-worker [34]. The text provides examples of several new materials for analytical proposes.

Table 1: Examples of current MXenes applications.

Application	Material	Description	Refs
Lithium ion batteries	Ti_2CT_x-EMD	Composite exhibit cycling stability improving Li-ion accessibility	[22]
Lithium-Sulphur batteries	Ti_3C_2 $Ti_3C_2O_x$	Usage of Ti_3C_2 and $Ti_3C_2O_x$ as sulfur supporting material for Li-S batteries	[23]
Sodium ion batteries	CoS/Ti_3C_2	Composite present better cycling stability and rate performance	[24]
Biological activity	Ti_2C-PEG	PEG coated MXene exhibit biocompatibility and photothermal conversion efficacy	[21]
Electrocatalyst	Pt/Ti_3C_2	Ti_3C_2 as support for Pt nanoparticles acting as electrocatalyst for methanol oxidation reaction and hydrogen evolution reaction	[25]
Microwave absorption	$Ti_3C_2T_x$/PANI	Polyaniline was polymerized in the surface and internal layers of MXenes, showing great potential in microwaves absorption	[20]
Optical communication	$Ti_3C_2T_x$	Integration in all-fiber optical system	[26]
Supercapacitor electrodes	$Ti_3C_2T_x$/Carbon fiber	MXene flakes were incorporated in carbon nanofibers, exhibiting high area capacitance as supercapacitor electrode	[27]
Supercapacitor	WO_3-Ti_3C_2	WO_3 nanoparticles evenly spread in MXene sheets granting enhanced capacitance	[28]
Antibacterial activity	Colloidal $Ti_3C_2T_x$	Colloidal nanosheets of $Ti_3C_2T_x$ exhibit antibacterial activity in bacteria (*B. subtilis and E. coli*)	[29]

EMD – Electrolytic Manganese Dioxide; PEG - Polyethylene glycol; PANI – Polyaniline.

As emphasized above and exemplified by Table 1, MXenes have a broad range of applications. The above mentioned works represent just a glimpse of the whole spectrum, and despite all the research and pragmatic applications – one does not fully understand how they really work. The aim of the present chapter resides in providing some recent examples of the development of two main classes of sensors, in which MXenes-base materials have been exploited: Electronic and biosensors. Recent examples dated from 2018 to 2019 are here described and discussed. A systematization of instrumental techniques usually employed in their characterization is also approached. In the following, before discussing these two sensor applications, some general aspects dealing with the synthesis of MXenes are presented.

2.   Synthesis of MXenes

Basically, there are three techniques to produce 2D materials: (i) exfoliation; (ii) physical vapour deposition (PVD) and (iii) chemical vapour deposition (CVD). There is a few exfoliation processes, such mechanical (scotch tape method), thermal, electrochemical and the most used the so-called the liquid phase exfoliation (LPE). In the LPE, a suitable solvent is used in order to overcome weak interlayer interactions and then to a sonication process where the exfoliation occurs [35]. The exfoliation process can be tuned to deliver selected isolated layers with control of shape, orientation, and thickness of the sheets. Nevertheless, this technique is limited to bulk layered materials.

In the PVD process, a solid material is vaporized under high temperature conditions or plasma under vacuum. The vapor is transported to the substrate surface where the condensation of the materials takes place as a film or 2D structure. In the case of CVD, the process is very similar to PDV, but the deposition process occurs in a reaction or decomposition of gas, liquid or solid materials on the substrate for the formation of the 2D materials [36]. The bottom-up synthesis approaches – as in the case of the PVD and CVD methods – the limitation of exfoliation process (bulk layered materials) is overcome, although these two approaches require total control of the synthesis conditions as ultrahigh vacuum and controlled atmosphere [4].

In spite of high-quality 2D ultrathin transition metal carbides (TMD) crystals have been reported employing CVD process [37], the exfoliation method is the most common technique used to synthesize MXenes. Historically, it was used by Naguib and collaborators [19] in the first report of MXenes synthesis back in 2011, in which titanium carbide (Ti_3C_2) through exfoliation of Ti3AlC2 using HF as etchant has been obtained.

In addition, especially in the case of the exfoliation method, usually MXenes are synthesized by the careful selection of an etchant - that normally contains fluoride (F^-), as

HF [19], NH_4HF_2, $NaHF_2$ or KHF_2 [38] – which is responsible for creating layers by removing *sp* elements from MAX phases. MAX correspondings to the formula $M_{n+1}AX_n$, where M corresponds to the *d*-block transition metal, such as Ti, Zr, Nb; A represents an *sp* element – more specific 13 and 14 groups, and X indicates either C and/or N [19,30,34]. Conversely to other two-dimensions materials where the interactions between layers are usually van der Waals forces, in MAX phases these interactions are normally metallic bonds [39]. Upon removing the element "A" through the etching process, the external surfaces layer present F, O and/or OH function groups. MXenes have a similar formula to that MAX phases, $M_{n+1}X_nT_x$ in which T represents the function group termination and x the number of terminations [30]. Figure 1 [40] illustrates the layered structured of MAX and the corresponding MXenes obtained by different treatments.

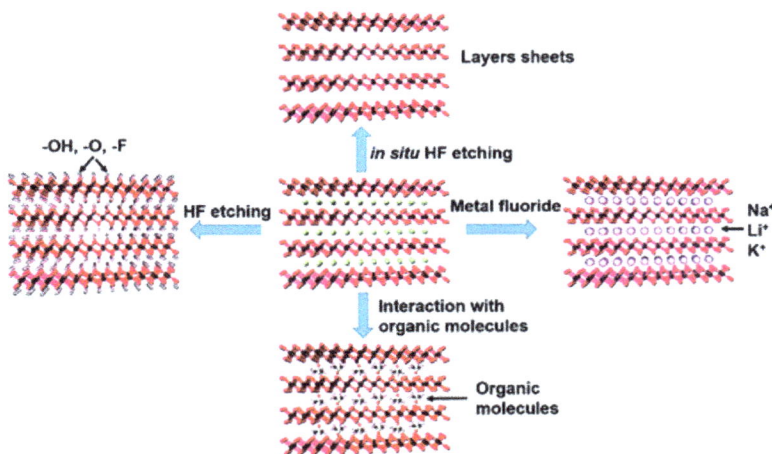

Figure 1: Schemetic of obtation of Mxenes from MAX trough different synthetic routes [40] – Reproduced by permission of The Royal Society of Chemistry.

As shown in Figure 1, after the replacement of the original metal bonding by van der Waals ones, a multilayer structure is formed (which again, will vary depending on the original MAX layer structure). Within the multilayered structure, there are molecules (or ions) that are intercalated depending on the nature of the solvent employed during the etching process: Water normally is associated when HF solution has been used, in the case of LiF/HCl, Li^+ also contributes to intercalation effects [41]. In fact, intercalation of organic molecules (as hydrazine, tetrabutylammonium, and urea) and other cations (like Na^+, K^+, NH_4^+ and more) have also been reported [33, 41, 42]. The volume of the intercalated molecule or ion reflects in the inter-layer spacing and thus changes some of its electronic and physical proprieties, especially the resistance. By annealing, for instance, it is possible to de-intercalate then, changing once again their properties, since the distance of the layer structure is shortened – resulting in a decreasing in resistance when the intercalation effects are present. Another outcome of the annealing process, after de-intercalation and under vacuum treatment, is the partial de-functionalization process on the surface, bringing different properties such higher conductivity in comparison to a normally functionalized MXene surface [41].

Taking advantage of weak interactions between layers, it is possible to separate them in single or in a small number of layers, through the delamination processes of multilayered MXenes. There are a few methods to delaminate MXenes. The simplest one consists in sonicating after the HF etching process (which has low yields). Examples of more complex ones involve the use of alternative etching solutions (mentioned above) to overcome yields issues [33]. More details on synthesis conditions are reported on the studies of Gogosti group [43] were a series of methods and protocols for different types of synthesis and post treatments are described step by step. As well as the texts from Naguib et al [19] and other reviews [32, 33, 41, 42] are also highly recommended.

3.   MXenes for sensing applications

For the majority of applications, MXenes samples are usually employed as thin films, composed of MXenes flakes. As previously mentioned, due to etching processes, which occur in aqueous solutions, the surface of the sheets are terminated with oxygen and/or fluoride function groups – affording electronic conductivity, hydrophilicity and stability properties to MXenes which in turn make them distinguished from other two-dimension materials [41]. Depending on the termination on the surface, based on estimations of the first principles of density functional theory (DFT), MXenes sheets may exhibit different electronic character, i.e., metallic or semiconductor one. Ti_2C with oxygen termination – monolayer of Ti_2CO_2 – presents a semiconductor character, and it is predicted to have a remarkable sensibility towards NH_3 gas, both in terms of adsorption/desorption on the surface monolayer, therefore making it a promising candidate for NH_3 sensor [44]. DFT calculation also predicts the same behavior for other MXenes with M_2CO_2 structure (M being a variety of metals as Ti, Zr, Sc, and Hf). Simulations suggest that NH_3 can be chemisorbed on MXenes and the resulting interaction can be weekend by introducing electrons, which turn suggests the feasibility in developping reversible NH_3 sensor [45].

Despite all the potentialities, MXenes may also have a few disadvantages such as easiness of suffering oxidation in the presence of oxygen atmosphere, and, the trend to agglomerate and to precipitate – in presence of organic solvents [60]. In order to overcome these drawbacks, researchers have modified and functionalized the MXenes surfaces. There are basically two approaches: (i) An organic one – where normally a polymer, such as PANI [20] , PEG [21], Nafion [46] and other organic molecules, is added to modify the surface – and (ii) an inorganic one in which nanoparticles – as some oxides like TiO_2 [47] and graphene [48] among other materials [49] - are grafted on the surface of MXenes.

Taking into account that the majority of the MXenes properties are related to its electronic characteristics, in the following section, some examples of electronic sensors development are discussed.

3.1   Electronic sensors

MXenes possess high electronic conductivity as a result of transition metal carbides (or nitrides) in their core, enabling their use in a broad range of electronic purposes. For instance, Shankar et al. [50] combined Ti_2CT_x sheets, graphite powder, and silicone oil, obtaining a homogenous mixture that fit into a homemade electrode for determination of Adrenaline (AD). The MXenes-based electrode exhibited better performance if compared to the classical graphite composite paste electrode, they achieve low detection limit of 9.5 nM and a higher electron-transfer rate (i.e., it provided a faster response time). The electrode was capable to both simultaneously and distinguished analyse AD, ascorbic acid and serotonin.

Kim et al. [51] created a chemical gas sensor device based on $Ti_3C_2T_x$ film, which due to its functionalized surface (usually hydroxyl, oxygen and flourine) exhibited high sensibility towards hydrogen-bonding gases (such as ethanol, acetone and ammonia) reaching a LOD of 50 parts per billion (ppb) level for acetone at room temperature. The authors also compared the sensitivity to these gases in others 2D materials: MoS_2, BP (black Phosphorous) and reduced graphene oxide (rGO). The sensors that are fabricated with semi-conducting materials (MoS_2 and BP) showed good selectivity to ammonia gas; while those fabricated with conducting materials (rGO and $Ti_3C_2T_x$) demonstrated relativity low response to the gases. Graphene-based dispositive did not even show any noticeable selectivity. The MXenes sensor also displayed the best signal-to-noise ratio (SNR): two orders of magnitude higher if compared to these other investigated materials, since their metallic channeling provides lower noise when compared to semiconducting channels. This result shows a promising candidate for sensing volatile organic compounds (VOCs) in which a ppb level of detection sensitivity is necessary.

A rather peculiar alternative to constructing an integrated electronic device for temperature monitoring was developed by Zhao et al. [48] using a 3D printing technique. They integrated micro-supercapacitor (MSC) and tandem line-type temperature sensor (TLTS) by duplex printing. For the positive electrode, an ink was prepared by mixing Ti_2CT_x with single-walled carbon nanotubes (SWCNTs), and for the negative electrode, similarly, rGO was mixed with SWCNTs. The mix of one-dimension carbon nanotubes with two-dimension MXenes and graphene resulted in a temperature sensor with a high sensitivity of 1.2% per degree Celsius, the high areal specific capacitance of 30.76 $mF.cm^{-2}$ and energy density of 8.37 $\mu Wh.cm^{-2}$.

MXenes can be easily oxidized when exposed in an environment rich in oxygen or water. Therefore, they are usually kept under an inert atmosphere, such as argon for instance. In the oxidation process, TiO_2 nanocrystals are formed. This process that may be considered as degradation of the material may also be responsible to the photoresponse to UV radiation which is usually absent in oxidation-free MXenes. In a study in Ti_2CT_x-TiO_2 film, photocurrent decayed very slowly in the inert atmosphere, but it was dramatically accelerated when argon was switched to ambient air or oxygen. Recently, Chertopalov el al. [47] showed that the process is reversibly reveling a promising start for photosensing.

3.2   Biosensing

As stated before, one trend within research on MXenes consists of doping or functionalizing MXnes sheets with either organic or inorganic groups. For instance, Zhang et al. [52] constructed a self-standard ratiometric fluorescence resonance energy transfer (FRET) nanoprobe for quantitative detection of exosomes overlaying MXenes with Cy3-CD63 aptamer, reaching a detection limit of 1.4 x 10^3 mL^{-1} which in comparison with ELISA method is one thousand times lower. This nanoprobe was shown to be sensitive to different types of exosomes – suggesting a universal platform that could be tuned by changing the aptamer type to the desired target.

A solid-state electrochemiluminescence (ECL) sensor was developed by Fang et al. [46]: MXene/Nafion/Ru showed better conductivity and enhanced ruthenium adsorption into the electrode surface compared with only Nafion/Ru electrode. Using tripropylamine as ECL coreactant – the sensor presented a LOD of 5nM, and it was able to distinguish a single-nucleotide mismatch in a test using human urine. This result indicates that a promising tool can be achieved for analytical and bioanalytical purposes. Peng and et al. [53] explored the property of MXenes of quenching the fluorescence of dye-labeled single-stranded DNA (ssDNA). The study reveals that Ti_3C_2 nanosheets interact differently ways with ssDNA and double-stranded DNA (dsDNA): While these nanosheets nearly extinguish the fluorescence on ssDNA, upon the formation of the dsDNA, the fluorescence is intensified. Adding the enzyme exonuclease III to the mixture, which enhanced the fluorescence and thus improved the sensibility, the authors created a magnified fluorescent "on-off" sensor for HPV detection. The resulting device was capable to detect the presence of HPV in real samples with fast response (a few minutes) with a LOD of 100 pM.

Guo and collaborators [54] constructed a biodegradable wearable pressure sensor by integrating MXene impregnated in tissue paper between layers of polylatic acid (PLA) with interdigitated electrodes. The sensor exhibits a LOD of 10.2 Pa and time response of 11 ms, and it was able to detect tactile signals and spatial pressure distribution. Also, the transient pressure sensor was attached to the human skin and obtained biomonitoring data, showing a promising device for biosensing and healthcare segment.

Another strategy to develop a strain sensor was taken by Zhang et al. [55], creating MXenes/hydrogel composite by mixing a hydrogel (which consists of PVA, water and antidehydratation additives) and MXenes ($Ti_3C_2T_x$) nanosheets. The sensor exhibited stretch ability up to 3400 % of it is the original size, transcending the stretch obtained for the pure hydrogel. The authors suggest that the interactions between MXenes and hydrogel may be attributed to the increase in viscoelastic proprieties, creating a second crosslink with negative charges surfaces of Ti_3C_2Tx. The composite also exhibits self-heal ability and change in the resistance upon different types of deformation (tensile and compressive). These electromechanical responses allow the sensor to detect motions and speed on its surface, being sensible to handwriting, facial motions, and vocal signs.

A different approach on flexible sensor pressure was made by Li et al. [56], who created a flexible piezoresistive pressure sensor using a polyurethane (PU) sponge as a backbone to the sensor. The device was developed by alternately dip-coating the PU sponge in a positive charge solution of chitosan (CS) and a negative charge solution of $Ti_3C_2T_x$ sheets, resulting in interface MXenes@CS@PU flexible sensor. The sensor presented a fast 19 ms time response, LOD of 30 μN and it was able to detect different sets of motions. The sensor recorded distinct changes in resistance when attached to a finger. The sensor also exhibits a remarkable sensibility being able to distinguish different deformations done by several small insects – like a fly, snail, reptile and a spider – and even the pressure done by a droplet of water.

Another dip-coating strategy was proposed by Li and co-workers [57] creating an MXenes-textile-based piezoresistive pressure sensor. A cotton textile used as support was dip-coated on a Mxenes solution. The resulting coated cotton textile (MXene-textile) was placed in an integrated electrode, and the entire device was involved in a polyimide (PI) film, as shown in Figure 2 [57]. Upon deformation of the textile, the MXenes sheets were brought closer enhancing the sensitive of the system. The sensor exhibits high sensitive and fast response (26 ms), being able to recognize a diverse type of input like voice recognition, finger touch, pulse wave, and bending. Its simple manufacture and performance show that MXenes can be integrated into smart touch devices – such touch displays, wearables electronics, and human-machines interface.

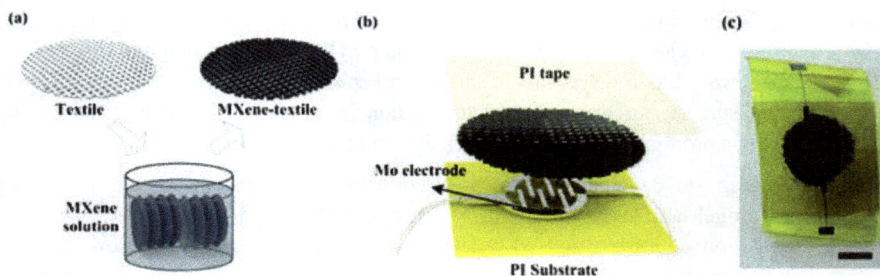

Figure 2: An schematic illustration of the fabrication process of the dispositive (a) Textile dip-coated in Mxene solution, obtaining MXene–textile. (b) MXene–textile based pressure sensor. (c) Photograph of the fabricated pressure sensor divice (scale bar, 5 mm). [57] – Reproduced by permission of The Royal Society of Chemistry.

Motjabavi et al. [58] synthesized two MXenes: $TiCT_x$ and $Ti_3C_2T_x$. Although both materials have the same synthetic route, the $TiCT_x$ have an additional intercalation process using tetra *n*-butyl ammonium hydroxide (TBAOH). They create an unusual approach in the preparation on MXene nanopore film, with a liquid-liquid interface, which after adding methanol in a chloroform-water interface which drives the MXenes sheets to the interfacial formation of the film. The MXene nanopores were able to sense double-stranded DNA molecules transport, obtaining similar results as to other two-dimension materials such as graphene. The $TiCT_x$ demonstrated better results if compared to $Ti_3C_2T_x$, apparently due to the surface functionalization: It bears more negative charge on the surface – and thus preventing the DNA transport. This result indicates that MXenes can point to a new direction for biosensing devices, once it is possible to obtain different results with MXenes characteristics such as hydrophilicity and conductivity.

Materials Research Forum LLC
doi: https://doi.org/10.21741/9781644900253-1

4. Characterization

The characterization of MXenes provides information to feedback synthesis approach, as well as to improve sensor performance. A set of complementary instrumental techniques is employed aiming to provide elemental, structural, textural and morphological information. The most employed technique is X-ray diffraction (XRD) that provides information about the difference in the crystal structure of MXenes, and their precursors allow monitoring the synthesis of MXenes through MAX phases. The most common MAX (Ti_3AlC_2) phase and the $Ti_3C_2T_x$ MXene will be taken as an example. The Ti_3AlC_2 has a XRD (002) peak at 9.5° (this technique can also be used to evaluate impurities in the MAX phase, the most common being Ti_3AlC where the peak is around 13° [43]) after the treatment with the etching solution, this (002) peak normally losses its intensity, broadens and shifts to lower angles. These changes in the peak are associated with the removal of Al atom and with the appearance of the functional groups (-OH, -O and –F) on the surface of MXenes, as can be observed in the schematic representation of the surfaces terminations in Figure 3a [59]. Another case is the very intense peak approximately at 39° on Ti_3AlC_2, which disappears completely on MXenes XDR pattern [19,43], as illustrated in Figure 3d.

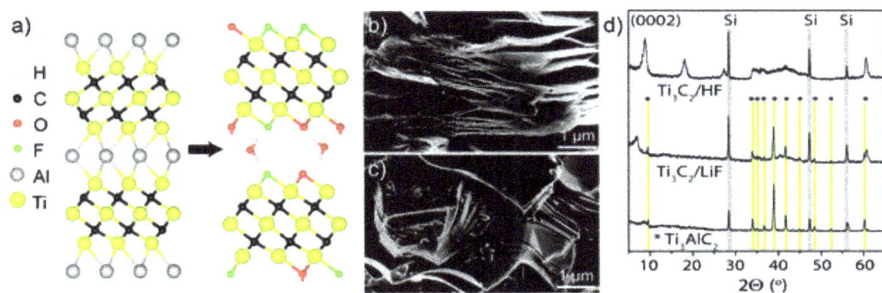

Figure 3: (a) Schematic of etching from Ti_3AlC_2 to $Ti_3C_2T_x$ showing terminations and interlayer water. Typical SEM micrographs of (b) HF-etched $Ti_3C_2T_x$, showing accordion-like structure and (c) LiF–HCl etched $Ti_3C_2T_x$, with a more compact structure. (d) XRD patterns of Ti_3AlC_2 and $Ti_3C_2T_x$ etched with HF or LiF–HCl. Green vertical lines denote the presence of MAX phase peaks, and grey regions are peaks from crystalline Si 10 wt% internal standard. [59] – Reproduced by permission of The Royal Society of Chemistry.

X-ray photoelectron spectroscopy (XPS) can contribute to determinate the functionalization of the surface by determining the atomic surface percentage and binding energy. For instance, by analyzing Ti 2p core level, it is possible to assign different binding energy (eV) to titanium species, resulting from the interactions as Ti-C Ti-X, Ti_xO_y, and TiO_2, for example (where Ti-X correspond to substoichiometric titanium carbides or oxycarbides). The same approach can be done for C, O and F 1s core levels [51]. A complementary technique to analyze the elemental composition of MXenes is the energy-dispersive X-ray spectroscopy (EDX or EDS), where it is possible to quantify the percentage of each element present [59]. Raman spectra before and after the etching process show different patterns. Taking Ti_3AlC_2 as an example, in lower Raman shifts (around 150 to 280 cm^{-1}) three peaks are observed that are associated with Ti-Al vibrations, that are complete vanish after the etching treatment. Three other peaks between 580 and 720 cm^{-1} are assigned to Ti-C vibrations, while in MAX phase there are three peaks that are merged, broadened and downshifted on MXenes [19].

Scanning electron microscopy (SEM) images provide morphological information about MAX phases and MXenes. MAX phases are bulk materials that normally show compact morphology. Conversely, MXenes, generally show an accordion-like structure, the morphology being accentuated depending on the nature of etchant. As a general rule, strong etchant as HF, both in higher and lower concentrations, impinges this accordion-like structure within few hours of reaction, while more mild approach of etching solutions does not afford this type of morphology, resulting in a structure in between the bulk compact MAX phase and vacancies like in MXenes [43], as can be observed in figure 3b and 3c.

Transmission electron microscopy (TEM) is a common technique to monitor the exfoliation of MXnes. Images normally reveal the very thin and transparent flake structure. In TEM it is possible to perform a crystallographic technique: Selected Area Electron Diffraction (SAD or SAED) which allows showing that MXenes have a hexagonal symmetry of the planes [19]. Atomic force microscopy (AFM) is applied to estimate the thickness of MXenes sheets, films, and composites. The thickness varies from a few nanometers to micrometers depending on the state of the sample [47, 52, 54]. Other techniques can be employed such as nuclear magnetic resonance (NMR) for the study of surface terminations [59] or cyclic voltammetry (CV) to evaluate electrochemical properties [48].

5.   Final Remarks

Due to their unique morphology aligned with their characteristic metallic conductive properties from the transition metal carbides/nitrides cores, MXenes have brought much attention both in the academic and scientific community in spite of its short time of existence, as emphasized in figure 4. The exponential growth of publication of documents and patents on MXenes theme in only eight years shows that MXenes holds a promising future in creating the next generations of composite materials technology.

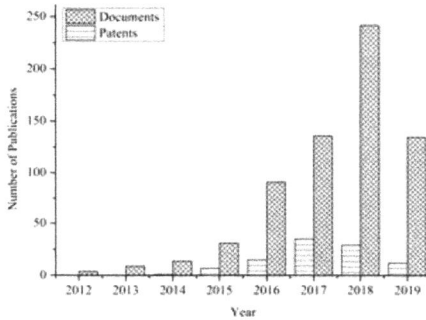

Figure 4: Graph representing the number of publications on the topic "MXenes" the years since 2011, in black are represented the documents published in the subject and in blue the patents on MXenes technologies. Source: Scopus search date: 04-12-2019.

In the field of sensor development, it has not been different: strategies in designing MXenes sensors have already consolidated their place in health, environment and medicine areas as illustrated in this chapter. Further applications can be found in other designs such as electroanalytical, photoluminescent and gas sensing [34]. Despite needing to deepen studies to better understand the electrochemical properties, MXenes bring a promising and exciting future to the development of sensor devices.

Acknowledgements

M.C. Cichero thanks CAPES for the grants. FAPERGS (16/2551-0000470-6) is also thanked.

References

[1]   F. Wang, Z. Wang, Q. Wang, F. Wang, L. Yin, K. Xu, Y. Huang, J. He, Synthesis, properties and applications of 2D non-graphene materials, Nanotechnology. 26 (2015) 292001. https://doi.org/10.1088/0957-4484/26/29/292001

[2]   D. Akinwande, C.J. Brennan, J.S. Bunch, P. Egberts, J.R. Felts, H. Gao, R. Huang, J.-S. Kim, T. Li, Y. Li, K.M. Liechti, N. Lu, H.S. Park, E.J. Reed, P. Wang, B.I.

Yakobson, T. Zhang, Y.-W. Zhang, Y. Zhou, Y. Zhu, A review on mechanics and mechanical properties of 2D materials—Graphene and beyond, Extrem. Mech. Lett. 13 (2017) 42–77. https://doi.org/10.1016/j.eml.2017.01.008

[3]   R. Mas-Ballesté, C. Gómez-Navarro, J. Gómez-Herrero, F. Zamora, 2D materials: to graphene and beyond, Nanoscale. 3 (2011) 20–30. https://doi.org/10.1039/C0NR00323A

[4]   A.J. Mannix, B. Kiraly, M.C. Hersam, N.P. Guisinger, Synthesis and chemistry of elemental 2D materials, Nat. Rev. Chem. 1 (2017) 14. https://doi.org/10.1038/s41570-016-0014

[5]   Y.L. and V.G. and C.J. and J.B. and A.L. and I.S. and A.P. and B.T. and P. Steyer, Advanced synthesis of highly crystallized hexagonal boron nitride by coupling polymer-derived ceramics and spark plasma sintering processes—influence of the crystallization promoter and sintering temperature, Nanotechnology. 30 (2019) 35604. http://stacks.iop.org/0957-4484/30/i=3/a=035604

[6]   X.W. and M.H. and Z.W. and L. Xie, Growth of two-dimensional materials on hexagonal boron nitride (h -BN), Nanotechnology. 30 (2019) 34003. http://stacks.iop.org/0957-4484/30/i=3/a=034003

[7]   W. Zheng, Y. Jiang, X. Hu, H. Li, Z. Zeng, X. Wang, A. Pan, Light Emission properties of 2D transition metal dichalcogenides: Fundamentals and applications, Adv. Opt. Mater. 6 (2018) 1800420. https://doi.org/10.1002/adom.201800420

[8]   J. Shi, M. Hong, Z. Zhang, Q. Ji, Y. Zhang, Physical properties and potential applications of two-dimensional metallic transition metal dichalcogenides, Coord. Chem. Rev. 376 (2018) 1–19. https://doi.org/10.1016/j.ccr.2018.07.019

[9]   A.K. Singh, P. Kumar, D.J. Late, A. Kumar, S. Patel, J. Singh, 2D layered transition metal dichalcogenides (MoS_2): Synthesis, applications and theoretical aspects, Appl. Mater. Today. 13 (2018) 242–270. https://doi.org/10.1016/j.apmt.2018.09.003

[10]   S. Yang, W. Hu, X. Zhang, P. He, B. Pattengale, C. Liu, M. Cendejas, I. Hermans, X. Zhang, J. Zhang, J. Huang, 2D covalent organic frameworks as intrinsic photocatalysts for visible light-driven CO_2 reduction, J. Am. Chem. Soc. 140 (2018) 14614–14618. https://doi.org/10.1021/jacs.8b09705

[11]   R. Mercado, R.-S. Fu, A. V Yakutovich, L. Talirz, M. Haranczyk, B. Smit, In silico design of 2D and 3D covalent organic frameworks for methane storage applications, Chem. Mater. 30 (2018) 5069–5086. https://doi.org/10.1021/acs.chemmater.8b01425

[12]   F.T. and E.P. and F.C. and F.C. and M.S.-R. and M.P. and S. Heun, Hybrid nanocomposites of 2D black phosphorus nanosheets encapsulated in PMMA polymer material: new platforms for advanced device fabrication, Nanotechnology. 29 (2018) 295601. http://stacks.iop.org/0957-4484/29/i=29/a=295601

[13]   H. Huang, Q. Xiao, J. Wang, X.-F. Yu, H. Wang, H. Zhang, P.K. Chu, Black phosphorus: a two-dimensional reductant for in situ nanofabrication, Npj 2D Mater. Appl. 1 (2017) 20. https://doi.org/10.1038/s41699-017-0022-6

[14]   Y. Yang, H. Hou, G. Zou, W. Shi, H. Shuai, J. Li, X. Ji, Electrochemical exfoliation of graphene-like two-dimensional nanomaterials, Nanoscale. 11 (2019) 16–33. https://doi.org/10.1039/C8NR08227H

[15]   R. Wang, X.-G. Ren, Z. Yan, L.-J. Jiang, W.E.I. Sha, G.-C. Shan, Graphene based functional devices: A short review, Front. Phys. 14 (2018) 13603. https://doi.org/10.1007/s11467-018-0859-y

[16]   H. Wang, T. Maiyalagan, X. Wang, Review on recent progress in nitrogen-doped graphene: Synthesis, characterization, and its potential applications, ACS Catal. 2 (2012) 781–794. https://doi.org/10.1021/cs200652y

[17]   Y. Zhang, L. Zhang, C. Zhou, review of chemical vapor deposition of graphene and related applications, Acc. Chem. Res. 46 (2013) 2329–2339. https://doi.org/10.1021/ar300203n

[18]   N.N. Rosli, M.A. Ibrahim, N. Ahmad Ludin, M.A. Mat Teridi, K. Sopian, A review of graphene based transparent conducting films for use in solar photovoltaic applications, Renew. Sustain. Energy Rev. 99 (2019) 83–99. https://doi.org/10.1016/j.rser.2018.09.011

[19]   M. Naguib, M. Kurtoglu, V. Presser, J. Lu, J. Niu, M. Heon, L. Hultman, Y. Gogotsi, M.W. Barsoum, Two-dimensional nanocrystals produced by exfoliation of Ti_3AlC_2, Adv. Mater. 23 (2011) 4248–4253. https://doi.org/10.1002/adma.201102306

[20]   H. Wei, J. Dong, X. Fang, W. Zheng, Y. Sun, Y. Qian, Z. Jiang, Y. Huang, Ti_3C_2Tx MXene/polyaniline (PANI) sandwich intercalation structure composites constructed for microwave absorption, Compos. Sci. Technol. 169 (2019) 52–59. https://doi.org/10.1016/j.compscitech.2018.10.016

[21]   A. Szuplewska, D. Kulpińska, A. Dybko, A.M. Jastrzębska, T. Wojciechowski, A. Rozmysłowska, M. Chudy, I. Grabowska-Jadach, W. Ziemkowska, Z. Brzózka, A. Olszyna, 2D Ti_2C (MXene) as a novel highly efficient and selective agent for photothermal therapy, Mater. Sci. Eng. C. 98 (2019) 874–886. https://doi.org/10.1016/j.msec.2019.01.021

[22]   S.A. Melchior, N. Palaniyandy, I. Sigalas, S.E. Iyuke, K.I. Ozoemena, Probing the electrochemistry of MXene (Ti_2CTx)/electrolytic manganese dioxide (EMD) composites as anode materials for lithium-ion batteries, Electrochim. Acta. 297 (2019) 961–973. https://doi.org/10.1016/j.electacta.2018.12.013

[23]   H. Pan, X. Huang, R. Zhang, D. Wang, Y. Chen, X. Duan, G. Wen, Titanium oxide- Ti_3C_2 hybrids as sulfur hosts in lithium-sulfur battery: Fast oxidation treatment and enhanced polysulfide adsorption ability, Chem. Eng. J. 358 (2019) 1253–1261. https://doi.org/10.1016/j.cej.2018.10.026

[24]   Y. Zhang, R. Zhan, Q. Xu, H. Liu, M. Tao, Y. Luo, S. Bao, C. Li, M. Xu, Circuit board-like CoS/MXene composite with superior performance for sodium storage, Chem. Eng. J. 357 (2019) 220–225. https://doi.org/10.1016/j.cej.2018.09.142

[25]   Y. Wang, J. Wang, G. Han, C. Du, Q. Deng, Y. Gao, G. Yin, Y. Song, Pt decorated Ti_3C_2 MXene for enhanced methanol oxidation reaction, Ceram. Int. 45 (2019) 2411–2417. https://doi.org/10.1016/j.ceramint.2018.10.160

[26]   Q. Wu, S. Chen, Y. Wang, L. Wu, X. Jiang, F. Zhang, X. Jin, Q. Jiang, Z. Zheng, J. Li, M. Zhang, H. Zhang, MZI-based all-optical modulator using MXene Ti_3C_2Tx (T = F, O, or OH) deposited microfiber, Adv. Mater. Technol. 0 (2019) 1800532. https://doi.org/10.1002/admt.201800532

[27]   A.S. Levitt, M. Alhabeb, C.B. Hatter, A. Sarycheva, G. Dion, Y. Gogotsi, Electrospun MXene/carbon nanofibers as supercapacitor electrodes, J. Mater. Chem. A. 7 (2019) 269–277. https://doi.org/10.1039/C8TA09810G

[28]   S.B. Ambade, R.B. Ambade, W. Eom, S.H. Noh, S.H. Kim, T.H. Han, 2D Ti_3C_2 MXene/WO_3 hybrid architectures for high-rate supercapacitors, Adv. Mater. Interfaces. 5 (2018) 1801361. https://doi.org/10.1002/admi.201801361

[29]   A. Arabi Shamsabadi, M. Sharifian Gh., B. Anasori, M. Soroush, Antimicrobial mode-of-action of colloidal Ti_3C_2Tx MXene nanosheets, ACS Sustain. Chem. Eng. 6 (2018) 16586–16596. https://doi.org/10.1021/acssuschemeng.8b03823

[30]   X. Li, C. Wang, Y. Cao, G. Wang, Functional MXene materials: Progress of their applications, Chem. – An Asian J. 13 (2018) 2742–2757. https://doi.org/10.1002/asia.201800543

[31]   K. Hantanasirisakul, Y. Gogotsi, Electronic and optical properties of 2D transition metal carbides and nitrides (MXenes), Adv. Mater. 0 (2018) 1804779. https://doi.org/10.1002/adma.201804779

[32]   B. Anasori, M.R. Lukatskaya, Y. Gogotsi, 2D metal carbides and nitrides (MXenes) for energy storage, Nat. Rev. Mater. 2 (2017) 16098. https://doi.org/10.1038/natrevmats.2016.98

[33]   J. Zhu, E. Ha, G. Zhao, Y. Zhou, D. Huang, G. Yue, L. Hu, N. Sun, Y. Wang, L.Y.S. Lee, C. Xu, K.-Y. Wong, D. Astruc, P. Zhao, Recent advance in MXenes: A promising 2D material for catalysis, sensor and chemical adsorption, Coord. Chem. Rev. 352 (2017) 306–327. https://doi.org/10.1016/j.ccr.2017.09.012

[34]   A. Sinha, Dhanjai, H. Zhao, Y. Huang, X. Lu, J. Chen, R. Jain, MXene: An emerging material for sensing and biosensing, TrAC Trends Anal. Chem. 105 (2018) 424–435. https://doi.org/10.1016/j.trac.2018.05.021

[35]   C. Backes, T.M. Higgins, A. Kelly, C. Boland, A. Harvey, D. Hanlon, J.N. Coleman, Guidelines for exfoliation, characterization and processing of layered materials produced by liquid exfoliation, Chem. Mater. 29 (2017) 243–255. https://doi.org/10.1021/acs.chemmater.6b03335

[36]   I.Y. Konyashin, PVD/CVD technology for coating cemented carbides, Surf. Coatings Technol. 71 (1995) 277–283. https://doi.org/10.1016/0257-8972(94)02325-K

[37]   C. Xu, L. Wang, Z. Liu, L. Chen, J. Guo, N. Kang, X.-L. Ma, H.-M. Cheng, W. Ren, Large-area high-quality 2D ultrathin Mo_2C superconducting crystals, Nat. Mater. 14 (2015) 1135. https://doi.org/10.1038/nmat4374

[38]   A. Feng, Y. Yu, Y. Wang, F. Jiang, Y. Yu, L. Mi, L. Song, Two-dimensional MXene Ti_3C_2 produced by exfoliation of Ti_3AlC_2, Mater. Des. 114 (2017) 161–166. https://doi.org/10.1016/j.matdes.2016.10.053

[39]   P. Srivastava, A. Mishra, H. Mizuseki, K.-R. Lee, A.K. Singh, mechanistic insight into the chemical exfoliation and functionalization of Ti_3C_2 MXene, ACS Appl. Mater. Interfaces. 8 (2016) 24256–24264. https://doi.org/10.1021/acsami.6b08413

[40]   N.K. Chaudhari, H. Jin, B. Kim, D. San Baek, S.H. Joo, K. Lee, MXene: an emerging two-dimensional material for future energy conversion and storage applications, J. Mater. Chem. A. 5 (2017) 24564–24579. https://doi.org/10.1039/C7TA09094C

[41]   J.L. Hart, K. Hantanasirisakul, A.C. Lang, B. Anasori, D. Pinto, Y. Pivak, J.T. van Omme, S.J. May, Y. Gogotsi, M.L. Taheri, Control of MXenes' electronic properties through termination and intercalation, Nat. Commun. 10 (2019) 522. https://doi.org/10.1038/s41467-018-08169-8

[42]   O. Mashtalir, M. Naguib, V.N. Mochalin, Y. Dall'Agnese, M. Heon, M.W. Barsoum, Y. Gogotsi, Intercalation and delamination of layered carbides and carbonitrides, Nat. Commun. 4 (2013) 1716. https://doi.org/10.1038/ncomms2664

[43]   M. Alhabeb, K. Maleski, B. Anasori, P. Lelyukh, L. Clark, S. Sin, Y. Gogotsi, Guidelines for synthesis and processing of two-dimensional titanium carbide (Ti_3C_2Tx

MXene), Chem. Mater. 29 (2017) 7633–7644.
https://doi.org/10.1021/acs.chemmater.7b02847

[44]   X. Yu, Y. Li, J. Cheng, Z. Liu, Q. Li, W. Li, X. Yang, B. Xiao, Monolayer Ti_2CO_2: A promising candidate for NH_3 sensor or capturer with high sensitivity and selectivity, ACS Appl. Mater. Interfaces. 7 (2015) 13707–13713.
https://doi.org/10.1021/acsami.5b03737

[45]   B. Xiao, Y. Li, X. Yu, J. Cheng, MXenes: Reusable materials for NH_3 sensor or capturer by controlling the charge injection, Sensors Actuators B Chem. 235 (2016) 103–109. https://doi.org/10.1016/j.snb.2016.05.062

[46]   Y. Fang, X. Yang, T. Chen, G. Xu, M. Liu, J. Liu, Y. Xu, Two-dimensional titanium carbide (MXene)-based solid-state electrochemiluminescent sensor for label-free single-nucleotide mismatch discrimination in human urine, Sensors Actuators B Chem. 263 (2018) 400–407. https://doi.org/10.1016/j.snb.2018.02.102

[47]   S. Chertopalov, V.N. Mochalin, Environment-sensitive photoresponse of spontaneously partially oxidized Ti_3C_2 MXene thin films, ACS Nano. 12 (2018) 6109–6116. https://doi.org/10.1021/acsnano.8b02379

[48]   J. Zhao, Y. Zhang, Y. Huang, X. Zhao, Y. Shi, J. Qu, C. Yang, J. Xie, J. Wang, L. Li, Q. Yan, S. Hou, C. Lu, X. Xu, Y. Yao, Duplex printing of all-in-one integrated electronic devices for temperature monitoring, J. Mater. Chem. A. 7 (2019) 972–978. https://doi.org/10.1039/C8TA09783F

[49]   H. Lin, Y. Chen, J. Shi, Insights into 2D MXenes for versatile biomedical applications: Current advances and challenges ahead, Adv. Sci. 5 (2018) 1800518. https://doi.org/10.1002/advs.201800518

[50]   S.S. Shankar, R.M. Shereema, R.B. Rakhi, Electrochemical determination of adrenaline using MXene/graphite composite paste electrodes, ACS Appl. Mater. Interfaces. 10 (2018) 43343–43351. https://doi.org/10.1021/acsami.8b11741

[51]   S.J. Kim, H.-J. Koh, C.E. Ren, O. Kwon, K. Maleski, S.-Y. Cho, B. Anasori, C.-K. Kim, Y.-K. Choi, J. Kim, Y. Gogotsi, H.-T. Jung, Metallic Ti_3C_2Tx MXene gas sensors with ultrahigh signal-to-noise ratio, ACS Nano. 12 (2018) 986–993. https://doi.org/10.1021/acsnano.7b07460

[52]   Q. Zhang, F. Wang, H. Zhang, Y. Zhang, M. Liu, Y. Liu, Universal Ti_3C_2 MXenes based self-standard ratiometric fluorescence resonance energy transfer platform for highly sensitive detection of exosomes, Anal. Chem. 90 (2018) 12737–12744. https://doi.org/10.1021/acs.analchem.8b03083

[53]   X. Peng, Y. Zhang, D. Lu, Y. Guo, S. Guo, Ultrathin Ti_3C_2 nanosheets based "off-on" fluorescent nanoprobe for rapid and sensitive detection of HPV infection, Sensors Actuators B Chem. 286 (2019) 222–229. https://doi.org/10.1016/j.snb.2019.01.158

[54]   Y. Guo, M. Zhong, Z. Fang, P. Wan, G. Yu, A wearable transient pressure sensor made with MXene nanosheets for sensitive broad-range human–machine interfacing, Nano Lett. (2019). https://doi.org/10.1021/acs.nanolett.8b04514

[55]   Y.-Z. Zhang, K.H. Lee, D.H. Anjum, R. Sougrat, Q. Jiang, H. Kim, H.N. Alshareef, MXenes stretch hydrogel sensor performance to new limits, Sci. Adv. 4 (2018) eaat0098. https://doi.org/10.1126/sciadv.aat0098

[56]   X.-P. Li, Y. Li, X. Li, D. Song, P. Min, C. Hu, H.-B. Zhang, N. Koratkar, Z.-Z. Yu, Highly sensitive, reliable and flexible piezoresistive pressure sensors featuring polyurethane sponge coated with MXene sheets, J. Colloid Interface Sci. 542 (2019) 54–62. https://doi.org/10.1016/j.jcis.2019.01.123

[57]   T. Li, L. Chen, X. Yang, X. Chen, Z. Zhang, T. Zhao, X. Li, J. Zhang, A flexible pressure sensor based on an MXene–textile network structure, J. Mater. Chem. C. 7 (2019) 1022–1027. https://doi.org/10.1039/C8TC04893B

[58]   M. Mojtabavi, A. VahidMohammadi, W. Liang, M. Beidaghi, M. Wanunu, Single-molecule sensing using nanopores in two-dimensional transition metal carbide (MXene) membranes, ACS Nano. (2019). https://doi.org/10.1021/acsnano.8b08017

[59]   M.A. Hope, A.C. Forse, K.J. Griffith, M.R. Lukatskaya, M. Ghidiu, Y. Gogotsi, C.P. Grey, NMR reveals the surface functionalisation of Ti_3C_2 MXene, Phys. Chem. Chem. Phys. 18 (2016) 5099–5102. https://doi.org/10.1039/C6CP00330C

[60]   K. Maleski, V.N. Mochalin, Y. Gogotsi, Dispersions of Two-Dimensional Titanium Carbide MXene in Organic Solvents, Chem. Mater. 29 (2017) 1632–1640. doi:10.1021/acs.chemmater.6b04830

Chapter 2

A Newly Emerging MXene Nanomaterial for Environmental Applications

Sze-Mun Lam[1,3,4,5]*, Ming-Wei Kee[1], Kok-Ann Wong[1], Zeeshan Haider Jaffari[1],
Huey-Yee Chai[1], Jin-Chung Sin[2,3,4,5]

[1]Department of Environmental Engineering, Faculty of Engineering and Green Technology,
Universiti Tunku Abdul Rahman, Jalan University, Bandar Barat, 31900 Kampar, Perak,
Malaysia

[2]Department of Petrochemical Engineering, Faculty of Engineering and Green Technology,
Universiti Tunku Abdul Rahman, Jalan University, Bandar Barat, 31900 Kampar, Perak,
Malaysia

[3]College of Environmental Science and Engineering, Guilin University of Technology, Guilin,
541004, China

[4]Guangxi Key Laboratory of Theory and Technology for Environmental Pollution Control,
Guilin University of Technology, Guilin 541004, China

[5]Collaborative Innovation Center for Water Pollution Control and Water Safety in Karst Area,
Guilin University of Technology, Guilin 541004, China

lamsm@utar.edu.my

Abstract

Escalating environmental issues have gathered prodigious scientific attention. Recently, two dimensional (2D) transition metal nitride/carbide composites have appeared as one of the most privileged nanomaterials applied in global environmental applications. Multilayered MXenes ($Ti_3C_2T_x$) were fabricated by exfoliating selective *MAX* phases. The description of the physiochemical properties of MXenes and their synthesis methods were detailed. This chapter also summarized the recent advancement of MXenes on environmental applications including adsorption, photocatalysis, antimicrobial and membrane processes. Finally, prospects together with challenges for MXenes possible environmental direction are summarized.

Keywords

MXene, Nanomaterial, Adsorption, Photocatalysis, Antimicrobial, Membrane

Contents

1.   Introduction

Intensifying environmental issue arose from rapid development has accumulated great attraction on the finding of feasible and sustainable technological solutions. The industrialization has resulted in a severe environmental problem which can inflict serious consequences on human health and the ecosystem. The general environmental contaminants comprised are heavy metals, organics, inorganics, bacteria, and poisonous gases. Multifarious nanomaterials have widely implemented in diverse applications including wastewater treatment [1-3], energy harvesting [4-8] and environmental sensing [9-10]. For instance, titanium dioxide (TiO_2) [11-13] and zinc oxide (ZnO) [14-16] have been widely studied for the destruction of organics from wastewater. Nonetheless, application of the aforementioned nanomaterials was technically restricted by some

MXenes: Fundamentals and Applications   Materials Research Forum LLC
Materials Research Foundations 51 (2019) 20-60   doi: https://doi.org/10.21741/9781644900253-2

practical obstacles. Firstly, their wide band gap structures represented that only UV part in the solar spectrum was utilized. Moreover, their rapid electron-hole pairs recombination minimized the provision of charge carriers adsorbed to the nanomaterials surface for the destruction of organics from wastewater.

Since the mechanical exfoliation of single layer graphene in the year 2004, two-dimensional (2D) materials have attracted significant interest due to their unique characteristics associated with their bulk counterparts. The isolation of single layer graphene has appeared a pioneer to all 2D nanomaterials and opened up the opportunity for more exploration. Ever since then, there are oodles of new 2D nanomaterials, such as hexagonal boron nitride, transition metal oxides, transition metal dichalcogenides (TMDs), and clays have been brought to light [17-22]. Recently, 2D metal-organic frameworks known as transition metal nitride/carbide composites (MXenes and MXene-based materials) has been employed in many applications, including ion batteries [17,18], triboelectric nanogenerators [19], supercapacitors [20-23], sensors [24,25], water purification and desalination [26-28]. The MXenes has garnered significant attention because of their remarkable physicochemical characteristics including huge surface area, outstandingly electrical and thermally conductive, high melting point, extraordinarily stable, hydrophilic and remarkably oxidation resistance.

In 2011, a simple chemical etching method was first used to fabricate MXenes ($Ti_3C_2T_x$), where an A layer was etching from a MAX matrix phase (Ti_3AlC_2). The MAX matrix phase has been termed as $M_{n+1}AX_n$ ($n = 1, 2, 3$), in which M signifies the transition metals (Cr, Sr, Ta, V, Ti, Zr, Nb, Hf or Mo), A denotes as the sp elements (generally IIIA or IVA), and X represents either N or C as well as both. The MXenes have been named from the general formula of $M_{n+1}X_nT_x$, where T_x indicates the functional termination groups (mostly –O, –OH and –F) [26-29]. The MXenes were fabricated by the wet-chemical etching method using hydrofluoric acid (HF) or HF-containing or HF-forming etchants [29,30] that provide surface functionalization with surface functional groups, including –O, –OH and –F. These surface functionalities enabled fast electronic and ion transport that in turn affected the conductivity and electron transfer processes on the surfaces of the MXenes. The fabrication condition of MXenes such as etching and delamination condition resulted in different surface functionalization properties and therefore correlated to the MXenes performance in their application.

The MXene materials have found to be next generation adsorbents and membranes to remove environmental contaminants because of their hydrophilic nature that enhanced the water flux [31-34]. Also, the abundancy of reactive functional sites on their surface have made the MXene materials to be efficient candidates for the removal of environmental contaminants (molecular or ionic species). The MXene materials

MXenes: Fundamentals and Applications   Materials Research Forum LLC
Materials Research Foundations **51** (2019) 20-60   doi: https://doi.org/10.21741/9781644900253-2

manifested superior adsorption capacities as compared to those of conventional adsorbents, such as graphene-based materials and other carbon materials that are normally employed in water and wastewater treatment. This chapter is presented as follows: (i) physiochemical properties and synthesis of MXenes nanomaterials; (ii) MXenes for adsorptive environmental application; (iii) MXenes for photocatalytic performance; (iv) MXenes for antimicrobial activity, and (v) MXenes for membrane filtration in the environmental application.

2.   Physiochemical properties of MXenes nanomaterials

2.1   Crystal structure

The MXenes have attracted tremendous attention since the pioneering work performed by Naguib et al. [35].The MXenes have a common formula of $M_{n+1}X_nT_x$ (n = 1, 2, 3), whereas M is a transition metal, such as titanium (Ti), molybdenum (Mo), niobium (Nb) or vanadium (V), X represents carbon (C) and/or nitrogen (N), T stands for the surface functional groups (–O, –OH and –F) and X is the number of surface functional groups [36–38]. Fig. 1 illustrates the MXene structure is derived from the *MAX* phase of selective layers in the hexagonal structure with space group *P63/mmc* [39]. The M layers are packed with X atoms between the octahedral sites and interleaved with A atoms layers [39,40]. A strong covalent, metallic and ionic bonding is detected in M–X bond, while weakly bonded possessing pure metallic nature is observed in M–layers [41]. Subsequently, M–A bond would have a tendency to decompose at high temperatures to form $M_{n+1}X_n$ [42-44], which resulted in recrystallization and development of 3D $M_{n+1}X_n$ rocksalt-like structure as shown in Fig. 2 [40]. The bonding mode of MXenes is held together by partial ionic bond, which makes the MX layer very difficult to separate by ultrasonication, dispersion or mechanical exfoliation [45,46].

As the traditional methods such as ultrasonication or mechanical cleavage are hard to isolate the MX layer from MAX precursor and thus a selective etching method was used to prepare layered structure MXenes [35]. The formation of MXenes could be observed using X-ray diffraction (XRD) analysis. According to the research performed by Kumar et al. [47], the characteristic (002) peak in XRD pattern shifted from 9.7° to 7.5° signified an increase in d-spacing for Ti_3C_2-MXene layers and confirmed the formation of MXene. Similar XRD patterns were also observed in other studies to verify the synthesis of MXenes [48,49].

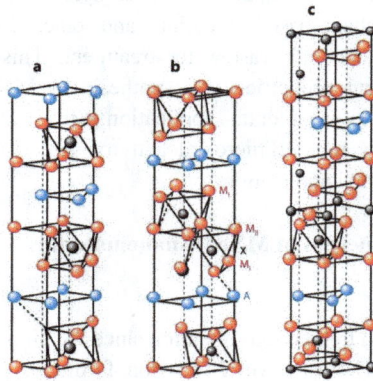

Figure 1. *Different types of MAX phase unit cells and crystal structures: (a) 211, (b)
312 and (c) 413 phases [39].*

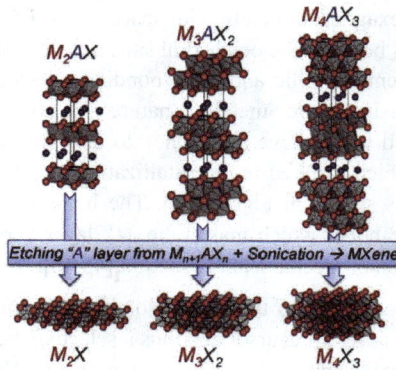

Figure 2. *Illustration of etching "A" layers from corresponding MAX phases [40].*

2.1.2 Surface chemical structure

Among the etching agents, hydrofluoric acid (HF) is widely applied for the etching
reaction as the bonding of *A* atoms are relatively weak. After the removal of *A*, the
exposed Ti become vigorously active and could easily functionalize with electronegative

charged as a ligand, which denoted as T_x in the general formula of $M_{n+1}X_nT_x$ [50,51]. This surface functionalization brings compelling effects on the ion and electron transport characteristics of MXenes, which could be straightly related to their conductivities and heterogeneous electron transport reaction taking place on the surfaces. For instance, Ti_3C_2 is a metal. Nevertheless, $Ti_3C_2OH_2$ or $Ti_3C_2F_2$ become semiconductors after surface functionalization. Thus, the surface chemical structure of MXenes could be modified by decorating these functional groups with different addends.

Generally, the surface functionalization of MXenes govern by the preparation conditions, including the delamination and etching conditions, enchant, post-synthetic procedure, M element, and storage type [52–54]. Moreover, numerous surface functional groups could alter MXenes characteristics significantly. Eames and Islam [55] stated that using O_2 as a surface functionalization tends to endorse the highest electrochemical performance as compared to those of –H or –OH groups functionalized MXenes. In another study, the bilayer $Ti_3C_2T_2$ with O-terminated has better ions intercalation than F-terminated one [56]. Furthermore, researchers have also studied other surface functionalization groups, such as P, Si and methoxy groups, which developed MXenes with enhanced properties [57,58].

2.1.3 Band gap structure

All the MXenes are metallic in nature owing to the presence of d orbitals of transition metal M. On the surface functionalization; several MXenes can exhibit semiconductor property with band gap energy ranged from around 0.05 to 1.80 eV, including Ti_2CO_2, Sc_2CO_2, Zr_2CO_2, Hf_2CO_2 and Mo_2CF_2 [59–64]. Table 1 shows the band gap energy of various MXenes at different surface functional groups [59-64]. According to work performed by Khazaei et al. [62], they stated that the presence of nearly free electron (NFE) states in various functionalized MXenes as displayed in Fig. 3. This could produce the dipole moments to allow the transfer of electron and cause semiconducting properties in the MXenes.

Based on the first-principle calculations, the electronic and magnetic properties of MXenes with low dimensionality differ momentously from those of their respective *MAX* phases in a suitable surface chemical treatment using various functionalization groups, like –OH, –F and –O [63, 65]. For instances, $Cr_2C(OH)_2$, Cr_2NF_2, $Cr_2N(OH)_2$ and Cr_2NO_2 have magnetic properties at low temperatures of around 100 K [63]. In another study, Shahzad et al. [66] stated that the synthesized $Ti_3C_2T_x$ MXene nanocomposites displayed supermagnetic attributes in the magnetic properties measurement system (MPMS) analysis. The saturation magnetization was 23.64 emu g^{-1} at room temperature in their studies. Also, MXenes also exhibited outstanding electromagnetic wave absorption property and could be utilized as light-to-heat conversion materials to generate heat for optical application [67,68].

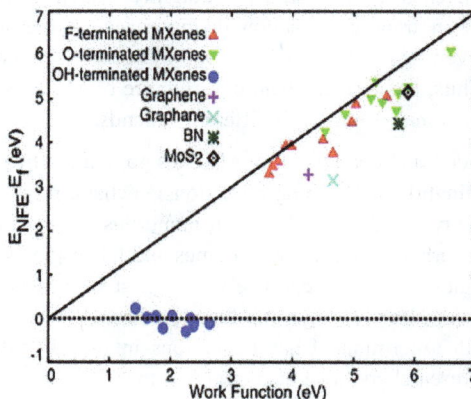

Figure 3.   The relative energy position of the lowest NFE state for functionalized MXenes, graphene, graphene, boron nitride (BN) and MoS_2 layers. The solid line represents the energy position of the vacuum level [62].

Table 1.   Band gap energy of various MXenes with different surface functional groups.

MXenes	Band gap energy [eV]	Type	Reference
Ti_2CO_2	0.24	Indirect band gap	[59]
$Ti_3C_2F_2$	0.10	-	[60]
$Ti_3C_2(OH)_2$	0.05	-	[60]
Sc_2CO_2	1.80	Indirect band gap	[61]
Sc_2CF_2	1.03	Indirect band gap	[62]
$Sc_2C(OH)_2$	0.45	Direct band gap	[62]
Zr_2CO_2	0.88	Indirect band gap	[63]
Hf_2CO_2	1.00	Indirect band gap	[63]
Mo_2CF_2	0.27	Indirect band gap	[64]

2.2   Synthesis of MXenes nanomaterials

Since the discovery of $Ti_3C_2T_x$ using HF etching of Ti_3AlC_2 in 2011 [35], the synthesis methods of MXenes have received great attention from the researchers. For the past few years, a variety of synthesis routes have been developed to prepare these MXenes. Table 2 [35,69-79] presents a summary of these synthesis methods.

Table 2.   *Summary of the preparation methods of MXenes.*

MXenes	Synthesis	Source	Temperature/time	Morphology	Surface area (m^2g^{-1})	Reference
Ti_3C_2	HF etching	Ti_3AlC_2	Room temperature/2 h	Fan	-	[35]
Ti_3C_2	HF etching	Ti_3AlC_2	60°C/24 h	Platelet	8.8	[69]
Mo_2ScC_2	HF etching	Mo_2ScAlC_2	50°C/16 h	-	-	[70]
$Nb_4C_3T_x$	HF etching	Nb_4AlC_3	Room temperature/140 h	Accordion	-	[71]
V_2C	HF etching	V_2AlC	Room temperature/144 h	Sheets	-	[72]
Ti_3C_2	$NaBF_4$–HCl etching	Ti_3AlC_2	180°C/35 h	Sheets	44.6	[74]
Ti_3C_2	LiF–HCl etching	Ti_3AlC_2	40°C/45 h	Flakes	-	[75]
Ti_3C_2	NH_4F–HCl etching	Ti_3AlC_2	30°C/24 h	Sheets	-	[76]
$Ti_3C_2T_x$	LiF–HCl etching	Ti_3AlC_2	35°C/24 h	Sheets	4.4188	[77]
$Ti_3C_2T_x$	Hydrothermal	Ti_3AlC_2	150°C/24 h	Sheets	47.25	[78]
Pt/Ti_3C_2	Hydrothermal	Ti_3AlC_2	80°C/2 h	Laminated	10.35	[79]
Ti_3C_2	Ultrasonic assisted etching	Ti_3AlC_2	25°C/4 h	Accordion	-	[73]

The MXenes synthesis was first revealed by Naguib et al. [35] in 2011. In their studies, Ti_3AlC_2 powders were placed in a HF solution (50 wt%) for 2 h under room temperature. After that, the resulting solution was washed many times using deionized water and centrifuged for particles separation. They also suggested a reaction mechanism to follow the Al atomic layers etching from the Ti_3AlC_2 layered phases. They stated that the selective etching was possible as the bonding of Ti-Al is more rapidly broken than Ti-C bonds as shown in Fig. 4 [35]. The reactions occurred when Ti_3AlC_2 was immersed in HF were as follow:

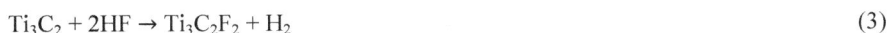

$$Ti_3AlC_2 + 3HF \rightarrow AlF_3 + 3/2\ H_2 + Ti_3C_2 \tag{1}$$

$$Ti_3C_2 + 2H_2O \rightarrow Ti_3C_2(OH)_2 + H_2 \tag{2}$$

$$Ti_3C_2 + 2HF \rightarrow Ti_3C_2F_2 + H_2 \tag{3}$$

MXenes: Fundamentals and Applications   Materials Research Forum LLC
Materials Research Foundations **51** (2019) 20-60   doi: https://doi.org/10.21741/9781644900253-2

The surface functional groups formation, including –F and –OH on MXenes, resulting in the 2 D Ti_3C_2 exfoliated layers surface terminations as depicted in Eq. 2 and 3. Hitherto, many MXenes have been synthesized using HF etching method [69–72].

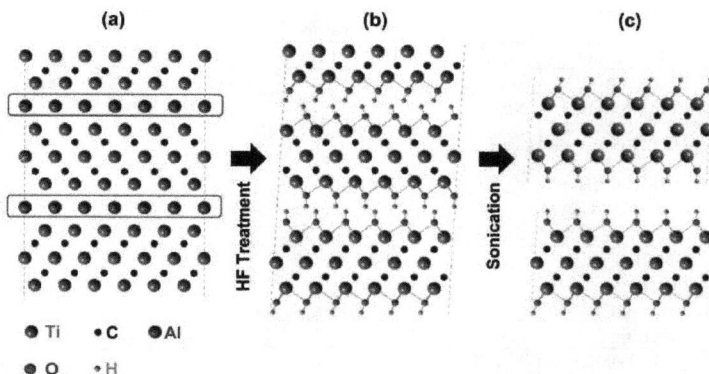

Figure 4.   Schematic illustration of the exfoliation process of Ti_3AlC_2, (a) Ti_3AlC_2 structure, (b) Al atoms being replaced by –OH after HF reaction, (c) Breakage of the hydrogen bonds and separation of nanosheets after sonication process [35].

Nevertheless, the usage of concentrated HF during the MXenes preparation restricted their applications due to the highly toxic and corrosive waste liquid generation [74, 80,]. Thus, it is important to propose an optional, efficient and harmless route for delamination and etching process. Peng et al. [74] developed an alternative route to etch Al from Ti_3AlC_2 using low toxicity etching agents like sodium tetrafluoroborate (NaBF$_4$) and hydrochloric acid (HCl) to synthesize Ti_3C_2 MXene. They stated that theTi_3C_2 synthesized by the alternative method has a higher c lattice parameter, larger interlayer distance and higher BET specific surface area as compared that of the Ti_3C_2 synthesized using the conventional HF etching technique. In another study, Xu et al. [77] also successfully synthesized $Ti_3C_2T_x$ by removing the Al element from Ti_3AlC_2 powders using an alternative etching route, which was verified by the XRD results (Fig. 5). They used the LiF–HCl etching process as an alternative to the HF etching process to eliminate the hazardous of the HF in the solution.

Numerous synthesis routes, including hydrothermal [78,79] and ultrasonic assisted etching [80] have also been revealed. A facile hydrothermal technique to synthesis $Ti_3C_2T_x$ by adding Ti_3AlC_2 powders in NH$_4$F solution and treated in an autoclave at

MXenes: Fundamentals and Applications   Materials Research Forum LLC

Materials Research Foundations **51** (2019) 20-60   doi: https://doi.org/10.21741/9781644900253-2

150°C for 24 h was reported by Wang et al. [78]. They mentioned that the hydrothermal route has avoided the use of HF in the solution, which makes this method safer and easier. Furthermore, Feng et al. [73] employed an ultrasonic assisted method for etching Ti_3C_2. They stated that this method could significantly reduce the HF etching time from 24 h to 4 h, which could simplify the synthesis method of MXene.

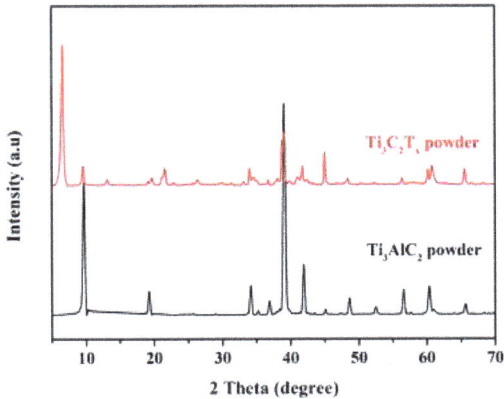

Figure 5.   XRD patterns of Ti_3AlC_2 and $Ti_3C_2T_x$ powder [77].

3.   MXenes for environmental application

3.1   Adsorption

A broad spectrum of aqueous/gaseous organic and inorganic pollutants has existed worldwide. Thus, efficient water and air purification treatment techniques are indispensable. Current traditional biological or physical purification methods are ranging from coagulation-flocculation, batch-operated aerobic biofilm system, rotating biological contactor reactor, the addition of powder/granular carbon, oxidation such as ozonation, chlorination and sonodegradation etc. [81–83]. Compared with all these techniques, adsorption is still favourable due to being feasible, low-cost and widely applicable [84]. As it is also well-known as carbon-based materials, MXenes could be delaminated into sheets, different organic pollutants, including phenol [85] and methyl orange [74] and heavy metals/metal ions, such asHg [66], Pb(II)[86], Cr(VI)[87], and Cd [88] indeed adsorbed by different MXenes. This signified that MXenes could be used as potential adsorbents. Hence, it is important to study the adsorption effects of MXenes towards the pollutants. The removal of aqueous/gaseous pollutants/heavy metal using MXene was influenced by the type of pollutants (e.g., organic/inorganic, charge and functional group), different adsorbent characterization (e.g., shape/size, functional group and surface area) and water/gas qualities (e.g., pH, temperature and adsorbate concentration).

3.1.1 Adsorption of organic pollutants

Organic pollutants are an important kind of aqueous contaminants, which typically have characteristics, such as resistance for biodegradation and chemical structure stability. Different dyestuffs, pesticides, phenol, and phenolic compounds are some of the major organic pollutants appeared in the wastewater. The existence of these organic pollutants in the wastewater has a direct impact on human health as well as on the entire ecosystem. Hence, the decomposition of organic pollutants is highly necessary. Some research groups have applied MXenes as adsorbents for the organic pollutant decomposition, and they reported some interesting findings.

The adsorption behaviour of MXene materials was different towards the charged state of dyes. Mashtalir et al. [31] synthesized $Ti_3C_2T_x$ MXene and utilized towards the removal of methylene blue (MB, cationic dye) and acid blue (AB80, anionic dye). Fig. 6[31] shows that $Ti_3C_2T_x$ exhibited an irreversible adsorption capacity of 39 mg/g towards the MB dye, while the AB80 dye was hardly removed. The higher adsorption capacity towards the cationic dye was mainly credited to the higher electrostatic attractions among the negatively charged MXene surface and positively charged cationic dye as compared to that of anionic dyes. Studies also proved that expending the interlayer spacing of MXenes were also helpful for adsorption efficiency. The interlayer spacing of $Ti_3C_2T_x$ was further enhanced by 29% and 28% via LiOH and NaOH treatment, respectively [89]. The NaOH-$Ti_3C_2T_x$ and LiOH-$Ti_3C_2T_x$ displayed the enhanced adsorption capacities of 189 mg/g and 121 mg/g, respectively towards the MB dye as compared that of 100 mg/g using pure$Ti_3C_2T_x$ MXene without alkaline treatment.

Wu et al. [85] synthesized $Ti_3C_2T_x$ MXene-based adsorbent materials and applied towards the removal of phenol from aqueous wastewater. The findings presented that the MXene showed good recycling, reproducibility and sustainable durability. The adsorption mechanism included the oxidation-reduction reaction of phenol. Meng et al., [90] synthesized $Ti_3C_2T_x$ MXene using a chemical synthesis method and successfully applied for the adsorption of urea from the wastewater and the residual obtained from the dialysis machine. The results showed that $Ti_3C_2T_x$ presented the lower adsorption capacity (10 mg/g) with the initial urea concentration (30 mg/dL) at room temperature. However, when the initial temperature was increased to the $37^{\circ}C$ more than two times higher adsorption capacity (21.7 mg/g) was recorded. The adsorption mechanism stated

MXenes: Fundamentals and Applications   Materials Research Forum LLC
Materials Research Foundations **51** (2019) 20-60   doi: https://doi.org/10.21741/9781644900253-2

that the molecules of urea easily entered the interlamellar space and formed hydrogen bonds with the hydrophilic groups presented on the MXene surface.

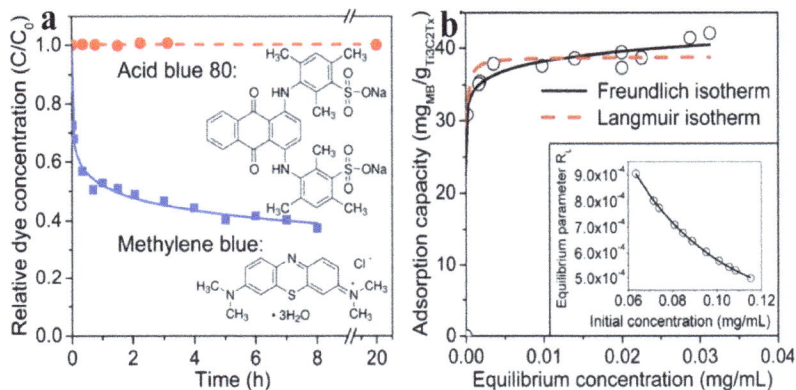

Figure 6.   (a) Time dependence of MB and AB 80 concentration in aqueous wastewater using $Ti_3C_2T_x$ MXene particles under dark conditions, (b) Adsorption isotherms of MB dye using $Ti_3C_2T_x$, insert of Figure shows the equilibrium parameters [31].

Zhang et al. [91] prepared O-terminated $Ti_3C_2O_2$ MXene nanosheet and theoretically studied for the removal of formaldehyde with the help of density functional theory (DFT)simulations. The results indicated that the $Ti_3C_2O_2$ MXene could adsorb formaldehyde molecules with an adsorption energy of 0.45 eV. The theoretically calculated adsorption capacity of $Ti_3C_2O_2$ towards the formaldehyde was 6 mmol/g. The adsorption mechanism proposed the existence of van der Waals forces among the H−O and C−O interactions. Collectively, these weak forces enabled the optimum adsorption energy, which was suitable for the adsorption of formaldehyde.

3.1.2 Adsorption of inorganic pollutants

The heavy metals and metal ions contaminants, including Hg, As, Cr(VI), Cd, Pb(II), Co and Cu exhibited a gigantic threat to the environment, particularly in the developing nations with less strict regulations [27,28]. Hence, the removal of these heavy metals/metal ions from aqueous solution were important for wildlife and human health. The adsorption process was the most economical as well as an effective approach for the removal of these heavy metal ions compared with the other available techniques. The MXenes with the abundant active sites can adsorb metal ions using chemical and electrostatic interactions. Until now, few studies have been performed on the adsorption of heavy metal ions using MXenes.

Lead ion Pb(II) was one of the generally found heavy metal ions in wastewater, and its presence in the drinking water is extremely harmful. Peng et al. [26] synthesized $Ti_3C_2(OH/ONa)_xF_{2-x}$ MXene using a chemical exfoliated method followed by alkalization interaction and first time applied for the adsorption of Pb(II). The results stated that the $Ti_3C_2(OH/ONa)_xF_{2-x}$ MXene presented higher adsorption capacity (140 mg/g) and great selectivity even in the presence of other competing cations, like Mg(II) and Ca(II). The intercalation of small organic molecules or cations could enhance the distance between interlayers of MXene, which led to the improvement in the interaction among the MXene layers and surface functional groups and thus improved the adsorption capacity. Particularly, after purification using the wastewater volume to adsorbent loading ratio of 4500 L/kg, the Pb(II) concentration reduced to lower than 2 mg/L, which was less than the standard set by the World Health Organization (WHO) (10 mg/L) for the drinking water.

The adsorption mechanism study revealed that the Pb(II) ions were trapped within the MXene layers and formed the strong bonds among the O atoms and Pb as displayed in Fig. 7 [27] . The results further described that the different functional groups and −OH sites also had a high influence on the Pb(II) ions adsorption capacities using $Ti_3C_2(OH)_xF_{2-x}$ MXenes. The vertical alignment of the −OH group to the Ti atom improved the adsorption of Pb(II) than those of the remaining structures, while the −F terminated occupation in alk-MXene reduced the removal efficiency and the intercalation of Na, K and Li atoms accelerate it. The DFT investigation of the Pb(II) adsorption using MXenes with the general formula of $M_2X(OH)_2$ (M = Cr, Sc, V, Ti, Mo, Nb, Ta, Hf,Zr, and X = N or C) showed that nitrides were mostly presented higher Pb(II) adsorption properties as compared to those of their carbide counterparts, which was because of the different numbers of electrons in the valence orbitals of C and N atoms [92].

Moreover, Cr(VI) is another hazardous heavy metal ions present in the aqueous phase, which exists in the $Cr_2O_7{}^{2-}$ form. The Cr(VI) adsorption was also investigated using MXene. Ying et al. [87] found the maximum removal of Cr(VI) (250 mg/g) on the $Ti_3C_2T_x$ MXene material. The remaining Cr(VI) concentration in the solution was lower than 5 mg. The MXene not only reduced the Cr(VI) to non-hazardous Cr(III) but also simultaneously adsorbed the reduced Cr(III). The electrostatic attraction among the negatively charged $Cr_2O_7{}^{2-}$ and positively charged MXene surface was the main reason for the Cr(VI) adsorption. After Cr(VI) adsorption, the electrons were transferred from the MXene to $Cr_2O_7{}^{2-}$ with the help of proton, which led to the production of Cr(III) ions and TiO_2. Afterwards, the generated Cr(III) ions interacted with the Ti−O bond of the

MXene and yielded the structure of Ti−O−Cr(III). Additionally, the other oxidizing agents like NaAuCl$_4$, K$_3$[Fe(CN)$_6$] and KMnO$_4$ were also effectively reduced to the lower oxidation states as well as effectively removed in the presence of Ti$_3$C$_2$T$_x$ MXene.

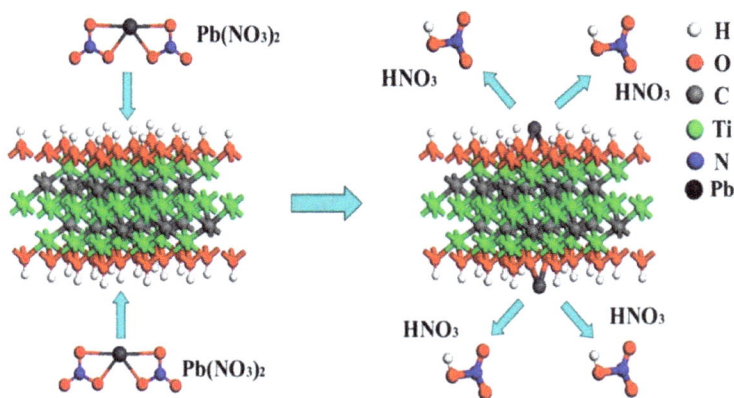

Figure 7.   *Schematic illustration of the formation of the Ti$_3$C$_2$(O$_2$H$_{2-2m}$Pb$_m$) structure from Ti$_3$C$_2$(OH)$_2$ to Ti$_3$C$_2$(O$_2$H$_{2-2m}$Pb$_m$) [27].*

In addition to the Pb(II) and Cr(VI) removals, Ti$_3$C$_2$ MXene was first time successfully applied for the decomposition of several other heavy metals, including Zn, Cu, Cd, and Pd from the wastewater [27]. The general reaction for the heavy metals (Y) ions adsorbed on the Ti$_3$C$_2$ MXene could be expressed as Ti$_3$C$_2$(OH)$_2$ + mY(NO$_3$)$_2$ → Ti$_3$C$_2$(O$_2$H$_{2-2m}$Y$_m$)$_2$ + 2mHNO$_3$. The estimated formation energies for Ti$_3$C$_2$(O$_2$H$_{2-2m}$Y$_m$)$_2$ were ranging from −1.0 to −3.3 eV for the Cd, Cu, Pd, and Zn. The Ti$_3$C$_2$T$_x$ MXenes were also successfully applied as adsorbents for Cu(II) [93] and Ba(II) ions [94]. The adsorption mechanism suggested that the Ba−F and Ba−O were mainly contributed in the adsorption of Ba(II) ions. On the other hand, the adsorption mechanism for Cu(II) ions experimentally confirmed the ion exchange among the negatively charged terminal –O and –OH groups and positively charged Cu(II) ions. Subsequently, oxidation-reduction reactions happened, which led to the production of rutile TiO$_2$ nanoparticles. In another study, Yang et al. [95] stated that the Ti$_3$C$_2$(OH)$_2$ and Ti$_2$C(OH)$_2$ MXenes could efficiently adsorb the free Au atoms with the adsorption energy ≥ 3 eV. The adsorption capacities of MXene and their derivatives for organic and inorganic contaminants were summarized in Table 3 [22,32,66,74,86,87,89-91,96-100].

Table 3. Summary of adsorptive removal of selected pollutants using MXene materials.

No	Adsorbent	Pollutant	Adsorbent dosage (g/L)	Initial concentration of pollutant	Time (h)	Temperature (°C)	pH	Adsorption capacity	Adsorption mechanism	Ref
1	$Ti_3C_2T_x$, LiOH–$Ti_3C_2T_x$, NaOH–$Ti_3C_2T_x$, KOH–$Ti_3C_2T_x$,	Methylene Blue	0.5	50 mg/L	3	25 25 25 25	6–6.5 8.8–9 8.8–9 8.8–9	100 mg/g 121 mg/g 189 mg/g 77 mg/g	Adsorption mechanism follows the Langmuir model, which indicates that alk-$Ti_3C_2T_x$ have homogeneous adsorption surfaces and only can have one adsorbate layer.	[89]
2	Ti_3C_2, Nb_2C	Methylene blue, Methyl orange	0.5	50 mg/L	3	25	–	24 mg_{MB}/g_{Ti3C2}, 5 mg_{MB}/g_{Ti3C2}, 21 $mg_{MO}/g_{Nb2C}3$, 3 $mg_{MO}/g_{Nb2C}3$	Electrostatic attractions among the positively charged MB dye and negatively charged MXene surface.	[74]
2	$BiFeO_3/Ti_3C_2T_x$	Congo red	1	20 mg/L	2	25	–	80 mg/g	Improved surface area caused the higher adsorption.	[96]
4	$Ti_3C_2T_x$, Ti_3CT_x, $Mo_2TiC_2T_x$	Urea	0.155	30 mg/dL	0.06 6 for $Ti_3C_2T_x$, 1 for	37	–	21.7 mg/g using $Ti_3C_2T_x$ 8.4 mg/g using $Mo_2TiC_2T_x$	Urea molecules easily enter the interlamellar space forming hydrogen bonds with the	[90]

					MoT iC$_2$T$_x$ and Ti$_3$C T$_x$			6.6 mg/g using Ti$_3$CT$_x$	hydrophilic surface functional groups.	[91]
5	Ti$_3$C$_2$O$_2$	Formaldehyde	0.05	5 mmol/L	–	0	–	6.9 mmol/g	Suitable adsorption energy of 0.45 eV caused the adsorption of formaldehyde.	[97]
6	Ti$_3$C$_2$T$_x$/Fe$_3$O$_4$	Real phosphate	0.5	10 mg/L	24	25	6.82	2400 kg/kg	Presence of strong specific attractions between the Fe-Ti hydroxide and phosphates.	
7	Ti$_3$C$_2$T$_x$	Hg(II)	0.025	25–1000 mg/L	24	42	6.0	1128.41mg/g	Electrostatic interaction between Ti–C–O$^-$ and Hg^{2+}	[66]
8	Ti$_3$C$_2$T$_x$	Cr(VI)	0.6	208 mg/L	72	27	5.0	250 mg/g	Electrostatic interaction between Ti–OH and Cr^{7+}	[87]
9	Ti$_3$C$_2$(OH)$_{0.8}$F$_{1.2}$	Cr(VI)	0.5	10 mg/L	12	25	6.0	225 mg/g	The bridging oxo-groups inhibited the H$_2$O adsorption, which led to the high Cr(VI) adsorption capacity.	[32]
10	Ti$_3$AlC$_2$	Pb(II)	0.2	70 mg/L	24	25	5.0	328.9 mg/g	Ion-exchange between Pb(II) and the interlayered	[86]

No.	Material		Concentration				Capacity	Mechanism	Ref.
11	Ti$_3$C$_2$(OH/ONa)$_x$F$_{2-x}$	Pb(II)	0.2	100 mg/L	0.5	50	6.5	140 mg/g	Na$^+$ ions was the key mechanism for Pb(II) removal. Ion-exchange between Pb(II) and –OH group on the MXene surface.	[26]
12	Alk-Ti$_3$C$_2$T$_x$	Ba(II)	0.5	50–500 mg/L	8	25	7–10	46.46 mg/g	Ion-exchange occurred between Ba ions and MXene.	[98]
13	V$_2$CT$_x$	U(VI)	0.4	5–120 mg/L	2	25	5	174 mg/g	Ion-exchange occurred between U(VI) and –OH group on the V site of MXene.	[99]
14	DL-Ti$_3$C$_2$T$_x$	Cu(II)	1	25 mg/L	0.16	45	5.0	78.45 mg/g	Ion-exchange between positively charged Cu ions and negatively charged terminal groups (–O and –OH) on the Ti$_3$C$_2$T$_x$ surface.	[93]
15	Ti$_3$C$_2$T$_x$	Bromate (BrO$_3^-$)	0.023	70 μmol/L	0.85	40	7.0	321.8 mg/g	The oxidation of Ti^{2+} to Ti^{4+} provides enough electrons to catalyze the reduction of BrO$_3^-$ into Br$^-$.	[100]

3.1.3 Adsorption of gaseous pollutants

Industrial discharged gaseous contaminates mostly consists of toxic inorganic gases such as NO_x, SO_x, H_2S, NH_3, CO_x and volatile organic compounds (VOCs). These gases can lead to severe infection to the human respiratory system and different other organs of the body. Some studies had been performed on the gaseous contaminants adsorption using MXenes.

Yu et al. [101] revealed the adsorption behaviors of different gases (H_2, NH_3, CO, CO_2, CH_4, NO_2 and O_2) on Ti_2CO_2 monolayer MXene. The findings indicated that only NH_3 gas was chemisorbed on the monolayer Ti_2CO_2 than those of the remaining gas molecules as shown in Fig. 8a. The N–Ti chemical interaction was the principle mechanism of the adsorption, while the calculated NH_3adsorption energy on Ti_2CO_2 was around -0.37 eV. The moderate energy value suggested that the Ti_2CO_2 was a promising recyclable adsorbent for NH_3 removal as it releases NH_3 easily. Moreover, the Ti_2CO_2 exhibited good improvement in the electrical conductivity after the adsorption of NH_3 gas. This result represented that Ti_2CO_2 can also be applied as a potential NH_3 sensor with high sensitivity as displayed in Fig. 8b. Morales-Garcia et al. [102]investigated the adsorption behaviour of pure M_2C (M = Cr,Hf, Mo, Ta, Ti, V, W, and Zr) MXenes towards the removal of CO_2 and found that the pure materials can efficiently adsorb the CO_2 even at lower partial pressure and higher temperature. Hence, they were very prominent candidates for CO_2 capture, storage, and activation. The mechanism for the adsorption involved complex interactions among the MXenes and CO_2 gas molecules and it also strongly dependent on the type of MXenes material. Since CO_2 was known to be chemically inert, these findings proved that the pure MXenes were extremely reactive for adsorbing the different pollutants.

Ma et al. [103]studied the adsorption capacity of SO_2 gas using monolayer O-terminated M_2CO_2 (M = Hf, Sc, Ti, and Zr). They found that compared with the other monolayers, Sc_2CO_2 was the most active adsorbent for SO_2 molecules due to the suitable adsorption strength (-0.646 eV). The main adsorption mechanism was the chemical bond between S–Sc. The adsorption capacity of Sc_2CO_2 can further improve by applying strains to the nanosheet. Moreover, the electric field had a strong effect on the SO_2 adsorption using Sc_2C. The negative electric field enhances the adsorption, while the positive electric field hinders it (direction from unadsorbed side to the adsorbed side). This distinctive property was really important for the applications of recyclable adsorbent materials and sensors. Liu et al. [76] synthesized Ti_3C_2 and Ti_2C MXene materials by etching Ti_2AlC and Ti_3AlC_2 along with different salts of fluoride such as KF, LiF, NH_4F, and NaF in HCl and applied towards the adsorption of CH_4 gas. All of the CH_4 adsorption isotherms showed improvement in the adsorption capacities with the increase of pressure. However, under the normal pressure, the MXenes exfoliated with LiF and NH_4F had the higher NH_4 adsorption capacities, whereas the MXenes exfoliated with KF and NaF only adsorbed CH_4 gas at a higher pressure and released it at the low pressure. Moreover, Ti_2C-MXenes had the highest adsorption capacities for CH_4 as compared to that of the Ti_3C-MXenes, which can be credited to their large surface areas.

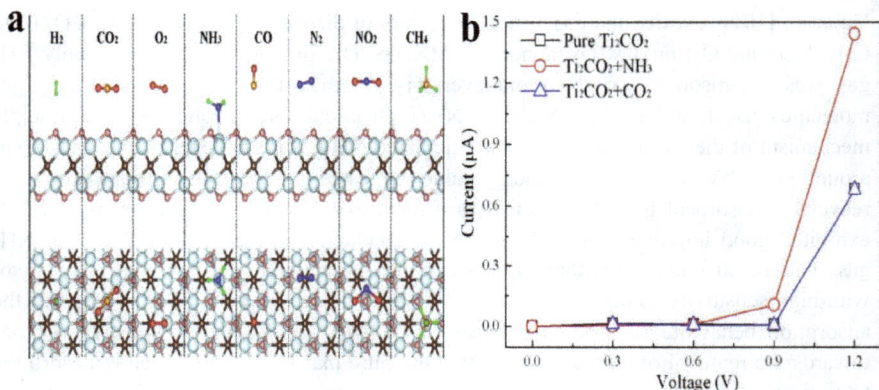

Figure 8.   (a) Schematic illustration of the adsorption on different gaseous such as NH_3,H_2, CH_4, CO, CO_2, N_2, NO_2 or O_2 molecule on Ti_2CO_2 MXene, (b) the current-voltage (I-V) relations before and after the adsorption of NH_3 or CO_2 molecule on monolayer Ti_2CO_2[96].

Apart from the adsorption of inorganic gaseous pollutants, MXenes were also applied as a sensor for VOCs. Kim et al. [25] applied $Ti_3C_2T_x$ MXene towards the detection of VOC gases including acetone, propanol, and ethanol. The results presented a very low detection limit (50−100 ppb) for VOC and NH_3 gases at the ambient temperature. The fully functionalize surface, and higher metallic conductivity of $Ti_3C_2T_x$ MXene resulted in an enhanced signal-to-noise ratio, which was greater than the conventional semiconductor materials along with other renowned 2D materials, for example, MoS_2. The terminal –OH group present on the $Ti_3C_2T_x$ surface was mainly accountable for the sensing of the gaseous contaminants.

MXenes: Fundamentals and Applications   Materials Research Forum LLC
Materials Research Foundations **51** (2019) 20-60   doi: https://doi.org/10.21741/9781644900253-2

3.1.4 Adsorption of other pollutants

The MXene materials have also exhibited excellent adsorption capacities towards several other hazardous pollutants because of their active surface structures. Zhang et al. [97] prepared sandwiched 2D MXene-iron oxide (*MXI*) via Al layer exfoliation and followed magnetic Fe_2O_3 intercalation for sequestration of phosphate in the wastewater. The MXI nanocomposite exhibited higher phosphate sequestration capacities of 2100 kg and 2400 kg/kg adsorption in simulated and real phosphate wastewater, respectively than those of the commercially available adsorbent $ArsenX^{np}$ (Purolite UK). The Fe_2O_3 ultrafine nanoparticles were intercalated in the inner layers of MXene, generating the Fe−OH exchanged sites and terminated Ti−OH layered surface inside the MXenes. The *MXI* materials were able to recycle using binary alkaline brine solution after phosphate adsorption.

Nuclear waste is another challenging concern for human health and the environment, which can promisingly remove by adsorption. Multilayered V_2CT_x MXenes were successfully applied as an adsorbent for actinide removal from the wastewater [99,104]. Uranium is a crucial nuclear material owing to its integral part in the nuclear industry. The V_2CT_x MXene can efficiently adsorb U(VI) with an extremely higher adsorption capacity (174 mg/g) accompanied by the desired selectivity and higher kinetics [99]. The X-ray absorption spectroscopy combined with the DFT calculation suggested that the interaction among the V_2CT_x and U(VI) followed the adsorption mechanism of ion-exchange, which was a typical chemisorption process. The $Ti_3C_2T_x$ was another promising MXene material, which showed adsorption properties towards the U(VI) removal [105]. However, the main challenge that restricted the adsorption capacity was the contradiction among the lower interlayer spacing in $Ti_3C_2T_x$ and the higher hydrated ionic radius of the U(VI), which caused a decline in the adsorption capacity. The adsorption capacity of $Ti_3C_2T_x$ material can significantly be enhanced by delamination of multilayered structures into nanoflakes. Additionally, hydration was another effective way to improve the adsorption characteristics of U(VI) ions using MXenes [106]. Using this technique, the existence of weak van der walls forces and hydrophilic groups can enhance the distance among the layers of MXenes, which led to the improvement in the adsorption capacity of U(VI) ions.

The MXene materials also had a remarkable application in the Li ion batteries, which was mainly due to their high adsorption capacity for Li ions. The Ti_3C_2 MXene was first time successfully applied for the Li ions adsorption with the capacity (320 mAh/g), which was equivalent to the 372 mAh/g using graphite towards the LiC_6[35]. Moreover, in the etching process, the −OH and –F groups were always terminated. Thus, their effects on the adsorption capacity of Li ions was also studied [107]. Briefly, the monolayer Ti_3C_2

Materials Research Forum LLC
doi: https://doi.org/10.21741/9781644900253-2

MXene showed a high Li ion adsorption capacity and a smaller Li ion diffusion barrier. On the contrary, the presence of −OH and –F groups on the surface of MXene hindered the transportation of Li ions and reduced the adsorption capacity. In another report, the effect of Li ion concentration and external strain on the Li ion adsorption using Ti_2C was also investigated [108]. The findings confirmed that the external strains showed a decline in bonding energy of Li atoms monotonically and it minutely depended on the Li concentration.

3.2 Photocatalysis

Photocatalytic technology is a promising approach to be extensively used to overcome environmental issue in recent years. It has garnered great attentions owing to its application of using semiconductors for complete degradation of recalcitrant organic pollutants into green products under suitable light irradiation, capable of being operated under ambient conditions including pressure and temperature, non-toxic and low operating cost [109,110]. Nowadays, there are various types of semiconductor available in the market for photocatalytic reaction, for example, metal oxide and MXene. Among them, MXene (known as transition metal carbides, nitrides and carbonitrides) is a newly emerged two-dimensional nanomaterial that has drawn massive researchers' attentions [111,112]. The MXene is a promising two-dimensional (2D) material with unique features, for instance, remarkable security capability, effective charge transfer, great biocompatibility, very large interlayer spacing and surface area, and environmental benignity [113,114]. The types of MXene as a candidate for photocatalytic applications including Sc_2CF_2, Ti_2CO_2, Hf_2CO_2, Zr_2CO_2, Sc_2CO_2 and etc. [40].

In a report, Zhang et al. [115] revealed their synthesized two-dimensional a-Fe_2O_3 nanosheets with layered Ti_3C_2 MXene performed 98% degradation of Rhodamine B under exposure of 2 h visible light. The excellent photocatalytic activity was attributed to numerous heterostructure interfaces on the MXene that allow strong absorption in the visible light region and great efficiency of charge carriers separation. Also, the synergistic effect also resulted in boosted photocatalytic activity compared to that of pristine Ti_3C_2. The significant result proved that the as-synthesized a-Fe_2O_3/Ti_3C_2 composite was suitable for photocatalytic activity towards organic pollutants degradation under exposure of visible light. Besides, Mashtalir et al. [31] fabricated $Ti_3C_2T_x$ on photocatalytic evaluation on acid blue 80 (AB80) and methylene blue (MB). After 5 h UV irradiation, the $Ti_3C_2T_x$ performed excellent degradation efficiency with 62 % AB80 and 81% MB, respectively.

Furthermore, Gao et al. [116] synthesized TiO_2/Ti_3C_2 nanocomposites through the hydrothermal process. The resulted material was examined on photocatalytic degradation of methyl orange (MO) in the presence of UV light irradiation. The TiO_2/Ti_3C_2 nanocomposites performed the excellent photocatalytic reaction with 98 % MO removal in 30 min. The efficient result was possibly attributed to the heterojunction between TiO_2 and Ti_3C_2 that allowed effective charge carriers separation as compared to that of pristine TiO_2 and Ti_3C_2.

A schematic diagram for potential photocatalytic mechanism using MXene materials $(Ti_3C_2/TiO_2/CuO$ nanocomposites) towards organic pollutants degradation is depicted in Fig. 9 [117]. Upon UV light irradiation, the photogenerated charge carriers are produced on both of the surfaces of TiO_2 and CuO. Due to the lower conduction band of CuO than TiO_2, the photoelectrons generated from TiO_2 will be transferred to the conduction band of CuO. Besides, both of the photogenerated electrons from TiO_2 and CuO will be transferred rapidly by Ti_3C_2 owing to the excellent electron conductivity to suppress the recombination rates of charge carriers. Therefore, the targeted pollutants can be degraded effectively by reactive oxidative species, including hydroxyl radicals and superoxide anions formed from a series of reduction and oxidation process into carbon dioxide and water.

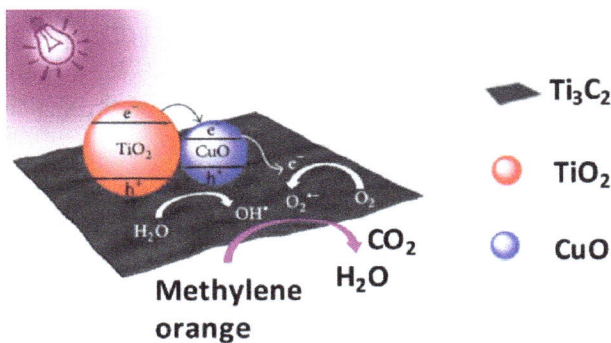

Figure 9.   *Schematic diagram of potential photocatalytic mechanism towards MO degradation [117].*

3.3   Antimicrobial activity

Based on the WHO, research and development on antibiotic resistance issues is always a top priority subject and therefore, continuous development of new antimicrobial agents is highly needed. In recent years, various 2D nanomaterials have gained significant interests in environmental and biomedical applications owing to their great potential in antimicrobial activity. An in-depth understanding of the primary antimicrobial mode of action of the 2D nanomaterials is vital for tailoring a new nanomaterial with higher antibacterial activity and lower toxicity towards human cells. Among a wide array of 2D nanomaterials, the graphene-based family including graphite, graphite oxide, graphene oxide, and reduced graphene oxide has been well known for their significant contribution in antibacterial activity [118,119]. The antibacterial activities over molybdenum disulfide (MoS_2) have also been well reported by several studies [120,121]. In contrast to these 2D nanomaterials, a newly emerged MXene was speculated that the sharp edge of MXene can cause damage to bacterial membranes and leading to bacterial death. Moreover, MXene also performs better in these contexts because of its outstanding physiochemical characteristics, including high oxidation resistance, hydrophilic, good biocompatibility, large interlayer surface area, and excellent stability, [122]. Hence, these characteristics render MXene as a promising candidate for antibacterial application.

In an effort to study the antibacterial potential of MXene, the interactions of MXene against Gram-positive and Gram-negative bacteria have been investigated. Rasool et al. [123] revealed the antibacterial activities of $Ti_3C_2T_x$ MXene flakes against *Escherichia coli* (*E. coli*) and *Bacillus subtilis* (*B. subtilis*). The results indicated that $Ti_3C_2T_x$ MXene flakes performed a superior antibacterial activity towards both Gram-negative *E. coli* and Gram-positive *B. subtilis* than the graphene oxide [123]. Arabi Shamsabadi et al. [124] reported the effects of size and exposure time of $Ti_3C_2T_x$ MXene nanosheets against both of *E. coli* and *B. subtilis*. Excellent antibacterial activities were observed against bothbacteria by smaller size nanosheets. The bacteria cells were damaged by the MXene nanosheets with the direct physical interactions between sharp edges of the nanosheets and membrane surfaces of bacteria within 3 h [124]. Besides, Mayerberger et al. [125] studied on the performance of crosslinked $Ti_3C_2T_z$ MXene flakes with chitosan nanofibers on antibacterial application. The cell reduction rate of about 95% and 62% of *E. coli* and *Staphylococcus aureus* (*S. aureus*) were reduced respectively, following a 4 h of treatment. This finding strongly suggests that MXene is a high biocompatibilities nanomaterial and display excellent antibacterial activity as a functionalized nanocomposite [125]. The recent studies of antibacterial activities using different MXenes were summarized in Table 4[121, 123-125].

MXenes: Fundamentals and Applications   Materials Research Forum LLC
Materials Research Foundations **51** (2019) 20-60   doi: https://doi.org/10.21741/9781644900253-2

Table 4.   *Recent studies of antibacterial activities using different MXenes.*

MXenes	Exposure time [h]	Type of targeted bacteria	Activity Performance[%]	References
$MoS_2/Ti_3C_2T_x$	3	*E. coli*	23	[121]
		B. subtilis	25	
$Ti_3C_2T_x$	4	*E. coli*	97	[123]
		B. subtilis	97	
$Ti_3C_2T_x$	3	*E. coli*	70	[124]
		B. subtilis	92	
$Ti_3C_2T_z$/chitosan	4	*E. coli*	95	[125]
		S. aureus	62	

It is crucial to study the interaction relationship between MXene with the bacteria cell membrane. The strong antibacterial activity of MXene can be ascribed to its surface absorption nature. Besides, a direct contact killing mechanism via the exposure of MXene edges with bacterial membrane surface was suggested. Furthermore, bacterial inhibition could happen through the H_2bonding between oxygenate groups of MXene and cell membrane lipopolysaccharide strings that can prevent nutrient intake [123]. From the contexts aforementioned, the antimicrobial mechanism of MXene can be proposed as shown in Fig.10 [126]. At first, the absorption ability of MXene surface edges allows them to effectively absorbed on the surface of microbes.

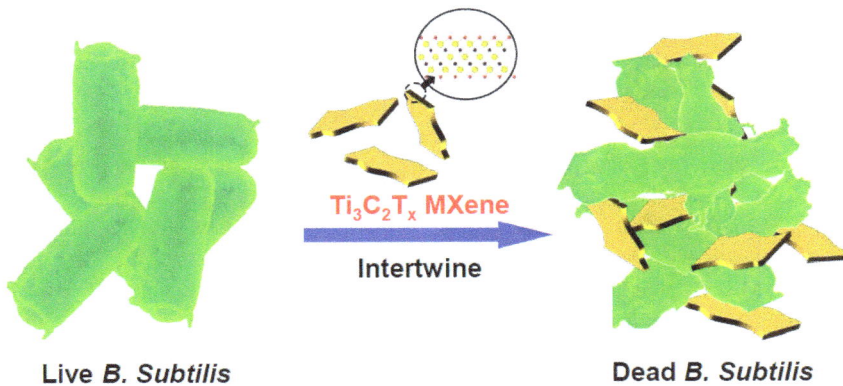

Figure 10.   *Schematics representation of the antibacterial activity of $Ti_3C_2T_x$ MXene [126].*

Consequently, the interaction between the sharp edges of MXene and microbes might induce membrane disruption. Subsequently, reactions in the cell wall and cytoplasm with MXene will take place and is likely to rupture the cell microstructure. Lastly, the severe membrane disruption and cell integrity leakage will eventually cause microbes death [126].

To date, there are still limited studies on the antibacterial property of MXene, mainly for environmental applications such as adsorption treatment, photocatalysis, and membrane biofouling agent [122]. However, the bigger picture is that to further explores the biological behavior and material science of 2D MXeneto facilitate the full potential use of 2D MXene in benefitting human health.

3.4   Membrane filtration

Over the years, membrane technology has become one of the potential methodologies to treat a diversity of water sources, which including nanofiltration, ultrafiltration, microfiltration and reverse osmosis [127]. These membrane filtration technologies offered several advantages such as excellent removal efficiency, ease cleaning process, the function of desalination, small carbon footprint contribution and less chemical requirement [122,128]. However, some drawbacks including membrane permeability and selectivity have caused the limitation of its application. To date, a lot of studies have been reported to against the limitation aforementioned by innovating the membrane materials with ultra-thin 2D materials, for instance, covalent organic nanosheets, metal organic frameworks, graphene based nanomaterials and MXenes [122, 129, 130]. Among them, MXenes have been used as material for the synthesis of an ultra-thin coating or separating layers to provide speedy passage of targeted solutions, in the meantime providing efficient permeability and remarkable selectivity to filtrate environmental pollutants [122].

There are some researches works have been conducted on the application of MXene membrane towards the removal of target pollutants. Liu et al. [131] synthesized ultrathin MXene ($Ti_3C_2T_x$) nanosheets membrane with ~60 nm by selective engraving followed by delamination process for application of pervaporation desalination. Their synthesized ultrathin MXene membrane possessed high hydrophilicity and unique 2D interlayer channels that allowed it performed excellent water flux at 85.4 L m^{-2} h^{-1} and also 99.5 % salt removal efficiency on 3.5 wt% NaCl solution. The product also exhibited effective long term stability performance and operation in synthetic seawater system. Besides, Han and his team [132] fabricated polyimide MXene mixed matric membrane via phase inversion method for gentian violet removal. Owing to the incorporation of inorganic additive, the as-fabricated membrane possessed remarkable layered structure as well as potential mass transfer channel that allowed it performed high water flux at 268 L m^{-2} h^{-1}

and 100 % removal on gentian violet. The membrane also exhibited excellent solvent resistance and perm selectivity, which is suggested to be employed in wastewater treatment with high salinity.

On the other hand, Xu et al. [77] also successfully synthesized 2D $Ti_3C_2T_x$ MXene incoporated into chitosan mixed matrix membrane by pervaporation process on the application of solvent deydration including dimethyl carbonate, ethanol, and ethyl acetate. The as-synthesized MXene laminates possessed multi-interlayer channels that improved the permeability of water molecule passing through the membrane. The high water flux at ~ 1.4-1.5 kg m^{-2} h^{-1} resulted in the membrane performed excellent selectivity for dehydration of dimethyl carbonate (906 selectivity), ethanol (1421 selectivity) and elthyl acetate (4898 selectivity). Compared to pristine chitosan membrane (water flux at 1.1-1.4 kg m^{-2} h^{-1}), the remarkable flux enhancement was attributed to the chitosan matrix the MXene membrane into layered that improved selective water molecule diffusion over another solvent. Pandey et al. [133] reported their synthesis of Ag modification on 2D nanostructured $Ti_3C_2T_x$ MXene (Ag@MXene) membrane for ultrafast water purification. The modified MXene membrane possessed superior water fluxthan that of pristine MXene membrane, which performed excellent organic molecules (bovine serum albumin and methyl green) rejection efficiency. Furthermore, the membrane also inhibited the growth of *Escherichia coli* (~99% inhibition) effectively compared to that of pristine MXene membrane with ~60 % growth inhibition of bacteria. Their modified Ag@MXene has garnered attentions as it demonstrated multifunctional purpose in the field of water treatment and biomedical owing to the combination of controllable permeability and bactericidal properties. Therefore, attributed to few of advantages from MXene including hydrophilicity, great ability of film forming, thermal and chemical stability, and antifouling properties, the membrane utilized MXene as a material has been explored and suggested to be used for future water purification method.

Conclusion and remarks

In this chapter, the physiochemical properties of MXenes materials have been described. Different surface functionalization is resulted from different synthesis conditions and can influence the performances of MXenes materials in their respective applications. The recent advancement of MXenes materials on the environmental applications, including adsorption, photocatalysis, antimicrobial and membrane have been presented. The studies showed that MXene had great potential for remediation of a broad spectrum of environmental pollutants including organic, inorganic and gaseous pollutants, probably owing to their different functional groups and huge surface areas. Other factors, including the nature of the pollutants, solution/gas chemistry and surface characteristics of MXene materials have a certain influence on the sorption process for aqueous/gaseous contaminants/heavy metals. MXene materials are superb substitutes relative to conventional adsorbents, including graphene-based materials and other carbon materials that are commonly used for water and wastewater treatment.

Photocatalytic technology using MXenes materials is a promising emerging solution for environmental remediation as it has shown exceptional performance to remove environmental pollutants. Its outstanding photocatalytic performance was attributed to its excellent electrical conductivity that enabled fast electron transfer and eventually facilitates the separation of charge carriers. MXenes materials have exhibited great antibacterial activity through direct contact killing mechanism. Direct contact of MXenes materials with bacteria could destroy the cellular membrane of bacteria that led to cell damage and consequent death. MXene-based membranes have shown extremely high capabilities for efficient destruction of organic and inorganic pollutants. Especially, incorporated 2D materials into a membrane structure have presented to substantially enhance membrane filtration process thru lower energy consumption, lower operating cost and a longer lifetime.

In spite of the achievement of substantial progress, there are still some vital technical points that necessary to be further inspected. They are described as follows: (i) lifecycle assessment of MXene materials required thorough research, (ii) long term stability as well as the adsorption abilities of MXene materials needed in-depth examination, (iii) mechanism of MXene materials required detailed studies ranging from experimental to computational methods, (iv) ecotoxicity of MXene materials needed comprehensive investigations. Taking into account on these points, the application of MXenes materials in environmental application would undeniably yield more stimulating outcomes in the near future.

Acknowledgments

The financial supports from Universiti Tunku Abdul Rahman (UTARRF/2018-C1/S02 and UTARRF/2018-C1/L02), Ministry of Higher Education of Malaysia (FRGS/1/2016/TK02/UTAR/02/1) and Research funds of The Guangxi Key Laboratory of Theory and Technology for Environmental Pollution Control (1801K012 and 1801K013) were gratefully acknowledged.

References

[1] M.W. Kee, J.W. Soo, S.M. Lam, J.C. Sin, A.R. Mohamed, Evaluation of photocatalytic fuel cell (PFC) for electricity production and simultaneous degradation

of methyl green in synthetic and real greywater effluents, J. Environ. Manage. 228 (2018) 383-392. https://doi.org/10.1016/j.jenvman.2018.09.038

[2]   K.A., Wong, S.M. Lam, J.C. Sin, Wet chemically synthesized ZnO structures for photodegradation of pre-treated palm oil mill effluent and antibacterial activity, Ceram. Int. 45 (2019) 1868-1880. https://doi.org/10.1016/j.ceramint.2018.10.078

[3]   S.W., Nam, C. Jung, H. Li, M. Yu, J.R.V. Flora, I. K. Boateng, N. Her, K.D. Zoh, Y. Yoon, Adsorption characteristics of diclofenac and sulfamethoxazole to graphene oxide in aqueous solution, Chemosphere 136 (2015) 20-26. https://doi.org/10.1016/j.chemosphere.2015.03.061

[4]   Y. Q., Zhu, C.B. Cao, A simple synthesis of two-dimensional ultrathin nickel cobaltite nanosheets for electrochemical lithium storage, Electrochim. Acta 176 (2015) 141-148. https://doi.org/10.1016/j.electacta.2015.06.130

[5]   Y.Q., Zhu, C.B. Cao, S. Tao, W.S., Chu, Z.Y. Wu, Y.D. Li, Ultrathin nickel hydroxide and oxide nanosheets: Synthesis, characterization and excellent supercapacitor performances, Sci. Rep. 4 (2014) 5787. https://doi.org/10.1038/srep05787

[6]   Y.Q., Zhu, C.B. Cao, J.T. Zhang, X.Y. Xu, Two-dimensional ultrathin $ZnCo_2O_4$ nanosheets: General formation and lithium storage application, J. Mater. Chem. A. 3 (2015) 9556-6564. https://doi.org/10.1039/c5ta00808e

[7]   Y.Q., Zhu, H.Z. Guo, H.Z. Zhai, C.B. Cao, Microwave-assisted and gram-scale synthesis of ultrathin SnO_2 nanosheets with enhanced lithium storage properties, ACS Appl. Mater. Interfaces 7 (2015) 2745-2753. https://doi.org/10.1021/am507826d

[8]   Y.Q., Zhu, W.M., Sun, W.X., Chen, T. Cao, Y. Xiong, J. Luo, J.C. Dong, I.R. Zheng, J. Zhang, X.I. Wang, C. Chen, Q. Peng, Scale-up biomass pathway to cobalt single-site catalysts anchored on N-doped porous carbon nanobelt with ultrahigh surface area, Adv. Funct. Mater. 28 (2018) 1802167. https://doi.org/10.1002/adfm.201802167

[9]   Q. Wang, X. F. Guo, L.C. Cai, Y. Cao, L. Gan, S. Liu, Z.X. Wang, H.T. Zhang, L.D. Li, TiO_2-decorated graphenes as efficient photoswitches with high oxygen sensitivity, Chem. Sci. 2 (2011) 1860-1864. https://doi.org/10.1039/c1sc00344e

[10]   Z. Bo, S. Mao, Z.J. Han, K.F. Cen, J.H. Chen, K. Ostrikov, Emerging energy and environmental applications of vertically-oriented graphenes, Chem. Soc. Rev. 44 (2015) 2108-2121. https://doi.org/10.1039/c4cs00352g

[11]   P. Sun, R. Xue, W. Zhang, I. Zada, Q. Liu, J. Gu, H. Su, Z. Zhang, J. Zhang, D.
Zhang, Photocatalyst of organic pollutants decomposition: TiO_2/glass fiber cloth
composites, Catal. Today 274 (2016) 2-7. https://doi.org/10.1016/j.cattod.2016.04.036

[12]   G. Song, Z. Chu, W. Jin, H. Sun, Enhanced performance of g-C_3N_4/TiO_2
photocatalysts for degradation of organic pollutants under visible light, Chin. J. Chem.
Eng. 23 (2015) 1326-1334. https://doi.org/10.1016/j.cjche.2015.05.003

[13]   K. Natarajan, H.C. Bajaj, R.J. Tayade, Effective removal of organic pollutants
using GeO_2/TiO_2 nanoparticle composites under direct sunlight, Mater. Chem. Front. 2
(2018) 741-751. https://doi.org/10.1039/c7qm00492c

[14]   D. Rajamanickam, M. Shanthi, Photocatalytic degradation of an organic pollutant
by zinc oxide – solar process, Arabian J. Chem. 9 (2016) S1858-S1868.
https://doi.org/10.1016/j.arabjc.2012.05.006

[15]   A. Gupta, J.R. Saurav, S. Bhattacharya, Solar light based degradation of organic
pollutants using ZnO nanobrushes for water filtration, RSC Adv. 5 (2015) 71472-
41481. https://doi.org/10.1039/c5ra10456d

[16]   P. Bansal, P. Kaur, D. Sud, 2014, Heterostructured TiO_2/ZnO–excellent
nanophotocatalysts for degradation of organic contaminants in aqueous solution,
Desalin. Water Treat. 52 (2014) 7004-7014.
https://doi.org/10.1080/19443994.2013.822330

[17]   M. Naguib, J. Halim, J. Lu, K.M. Cook, L. Hultman, Y. Gogotsi andM.W.
Barsoum, New two-dimensional niobium and vanadium carbides as promising
materials for Li-ion batteries, J. Am. Chem. Soc. 135 (2013) 15966-15969.
https://doi.org/10.1021/ja405735d

[18]   Y. Xie, M. Naguib, V.N. Mochalin, M.W. Barsoum, Y. Gogotsi, X. Yu, K.W.
Nam, X.Q. Yang, A.I. Kolesnikov and P.R. Kent, Role of surface structure on Li-ion
energy storage capacity of two-dimensional transition-metal carbides, J. Am. Chem.
Soc. 136 (2014) 6385-6394. https://doi.org/10.1021/ja501520b

[19]   Y. Dong, S.S.K. Mallineni, K. Maleski, H. Behlow, V.N. Mochalin, A.M. Rao, Y.
Gogotsi and R. Podila, Metallic MXenes: A new family of materials for flexible
triboelectric nanogenerators, Nano Energy 44 (2018) 103-110.
https://doi.org/10.1016/j.nanoen.2017.11.044

[20]   M. Ghidiu, M.R. Lukatskaya, M.Q. Zhao, Y. Gogotsi and M.W. Barsoum,
Conductive two-dimensional titanium carbide 'clay' with high volumetric
capacitance, Nature 516 (2014) 78. https://doi.org/10.1038/nature13970

[21]   J. Come, M. Naguib, P. Rozier, M.W. Barsoum, Y. Gogotsi, P.L. Taberna, M. Morcrette and P. Simon, A non-aqueous asymmetric cell with a Ti_2C-based two-dimensional negative electrode, J. Electrochem. Soc. 159 (2012) A1368-A1373. https://doi.org/10.1149/2.003208jes

[22]   M.R. Lukatskaya, S. Kota, Z. Lin, M.Q. Zhao, N. Shpigel, M.D. Levi, J. Halim, P.L. Taberna, M.W. Barsoum, P. Simon and Y. Gogotsi, Ultra-high-rate pseudocapacitive energy storage in two-dimensional transition metal carbides, Nat. Energy 2 (2017) 17105. https://doi.org/10.1038/nenergy.2017.105

[23]   C. Zhang, B. Anasori, A. Seral Ascaso, S.H. Park, N. McEvoy, A. Shmeliov, G.S. Duesberg, J.N. Coleman, Y. Gogotsi and V. Nicolosi, Transparent, flexible, and conductive 2D titanium carbide (MXene) films with high volumetric capacitance, Adv. Mater. 29 (2017) 1702678. https://doi.org/10.1002/adma.201702678

[24]   E. Lee, A. VahidMohammadi, B.C. Prorok, Y.S. Yoon, M. Beidaghi and D.J. Kim, Room temperature gas sensing of two-dimensional titanium carbide (MXene), ACS Appl. Mater. Interfaces 9 (2017) 37184-37190. https://doi.org/10.1021/acsami.7b11055

[25]   S.J. Kim, H.-J. Koh, C.E. Ren, O. Kwon, K. Maleski, S.-Y. Cho, B. Anasori, C.-K. Kim, Y.-K. Choi, J. Kim, Y. Gogotsi, H.-T. Jung, Metallic $Ti_3C_2T_x$ MXene gas sensors with ultrahigh signal-to-noise ratio, ACS Nano. 12 (2018) 986–993. https://doi.org/10.1021/acsnano.7b07460

[26]   Q. Peng, J. Guo, Q. Zhang, J. Xiang, B. Liu, A. Zhou, R. Liu, Y. Tian, Unique lead adsorption behavior of activated hydroxyl group in two-dimensional titanium carbide, J. Am. Chem. Soc. 136 (2014) 4113–4116. https://doi.org/10.1021/ja500506k

[27]   J. Guo, Q. Peng, H. Fu, G. Zou, Q. Zhang, Heavy-metal adsorption behavior of two-dimensional alkalization-intercalated MXene by first-principles calculations, J. Phys. Chem. C. 119 (2015) 20923–20930. https://doi.org/10.1021/acs.jpcc.5b05426

[28]   X. Guo, X. Zhang, S. Zhao, J. Xue, High adsorption capacity of heavy metals on two-dimensional MXenes: an ab initio study with molecular dynamics simulation, Phys. Chem. Chem. Phys. 18 (2016) 228–233. https://doi.org/10.1039/c5cp06078h

[29]   J. Zhou, X. Zha, X. Zhou, F. Chen, G. Gao, S. Wang, C. Shen, T. Chen, C. Zhi, P. Eklund and S. Du, Synthesis and electrochemical properties of two-dimensional hafnium carbide, ACS Nano 11 (2017) 3841-3850. https://doi.org/10.1021/acsnano.7b00030

[30]   B. Anasori, M.R. Lukatskaya and Y. Gogotsi, 2D metal carbides and nitrides (MXenes) for energy storage. Nat. Rev. Mater. 2 (2017) 16098. https://doi.org/10.1038/natrevmats.2016.98

[31]   O. Mashtalir, K.M. Cook, V.N. Mochalin, M. Crowe, M.W. Barsoum, Y. Gogotsi, Dye adsorption and decomposition on two-dimensional titanium carbide in aqueous media, J. Mater. Chem. A 2 (2014) 14334-14338. https://doi.org/10.1039/c4ta02638a

[32]   G.D. Zou, J.X. Guo, Q.M. Peng, A.G. Zhou, Q.R. Zhang, B.Z. Liu, Synthesis of urchin-like rutile titania carbon nanocomposites by iron-facilitated phase transformation of MXene for environmental remediation, J. Mater. Chem. A 4 (2016) 489-499. https://doi.org/10.1039/c5ta07343j

[33]   R.L. Han, X.F. Ma, Y.L. Xie, D. Teng, S.H. Zhang, Preparation of a new 2D MXene/PES composite membrane with excellent hydrophilicity and high flux, RSC. Adv. 7 (2017) 56204-56210. https://doi.org/10.1039/c7ra10318b

[34]   L. Ding, Y.Y. Wei, Y.J. Wang, H.B. Chen, J. Caro, H.H. Wang, A two-dimensional lamellar membrane: MXene nanosheet stacks, Angew. Chem. Int. Ed. 56 (2017) 1825-1829. https://doi.org/10.1002/anie.201609306

[35]   M. Naguib, M. Kurtoglu, V. Presser, J. Lu, J. Niu, M. Heon, L. Hultman, Y. Gogotsi, W. Barsoum, Two-dimensional nanocrystals produced by exfoliation of Ti_3AlC_2, Adv. Mater. 23 (2011) 4248 – 4253. https://doi.org/10.1002/adma.201102306

[36]   J. Peng, X. Chen, W. Ong, X. Zhao, N. Li, Surface and heterointerface engineering of 2D MXenes and their nanocomposites: insights into electro- and photocatalysis, Chem. 5 (2019) 18 – 50. https://doi.org/10.1016/j.chempr.2018.08.037

[37] S. Chen, Y. Xiang, C. Peng, J. Jiang, W. Xu, R. Wu, Photo-responsive Azobenzene-MXene hybrid and its optical modulated electrochemical effects, J. Power Sources 414 (2019) 192 – 200. https://doi.org/10.1016/j.jpowsour.2019.01.009

[38]   P. Liu, Z. Yao, V.M.H. Ng, J. Zhou, L.B. Kong, K. Yue, Facile synthesis of ultrasmall Fe_3O_4 nanoparticles on MXenes for high microwave absorption performance, Composites Part A 115 (2018) 371 – 382. https://doi.org/10.1016/j.compositesa.2018.10.014

[39]   J. Zhu, E. Ha, G. Zhao, Y. Zhou, D. Huang, G. Yue, L. Hu, N. Sun, Y. Wang, L.Y.S. Lee, C. Xu, K. Wong, D. Astruc, P. Zhao, Recent advance in MXenes: a promising 2D material for catalysis, sensor and chemical adsorption, Coord. Chem. Rev. 352 (2017) 306 – 327. https://doi.org/10.1016/j.ccr.2017.09.012

[40]   A. Sinha, Dhanjai, H. Zhao, Y. Huang, X. Lu, J. Chen, R. Jain, MXene: an emerging material for sensing and biosensing, Trends Anal. Chem. 105 (2018) 424 – 435. https://doi.org/10.1016/j.trac.2018.05.021

[41]   M. Naguib, V.N. Mochalin, M.W. Barsoum, Y. Gogotsi, 25th anniversary article: MXenes: a new family of two-dimensional materials, Adv. Mater. 26 (2014) 992 – 1005. https://doi.org/10.1002/adma.201304138

[42]   M. Naguib, O. Mashtalir, J. Carlet, V. Presser, J. Lu, L. Hultman, Y. Gogotsi, M.W. Barsoum, Two-dimensional transition metal carbides, ACS nano 6 (2012) 1322 – 1331. https://doi.org/10.1021/nn204153h

[43]   B. Anasori, Y. Xie, M. Beidaghi, J. Lu, B.C. Hosler, L. Hultman, P.R.C. Kent, Y. Gogotsi, M.W. Barsoum, Two-dimensional, ordered, double transition metals carbides (MXenes), ACS nano 9 (2015) 9507 – 9516. https://doi.org/10.1021/acsnano.5b03591

[44]   B. Anasori, C. Shi, E.J. Moon, Y. Xie, C.A. Voigt, P.R.C.Kent, S.J. May, S.J.L. Bilinge, M.W. Barsoum, Y. Gogotsi, Control of electronic properties of 2D carbides (MXenes) by manipulating their transition metal layers, Nanoscale Horiz. 1 (2016) 227 – 234. https://doi.org/10.1039/c5nh00125k

[45]   Q. Tang, Z. Zhou, Graphene-analogous low-dimensional materials, Prog. Mater. Sci. 58 (2013) 1244 – 1315.

[46]   K.S. Novoselov, D. Jiang, F. Schedin, T.J. Booth, V.V. Khotkevich, S.V. Morozov, A.K. Geim, Two-dimensional atomic crystals, Proc. Natl. Acad. Sci. 102 (2005) 10451 – 10453. https://doi.org/10.1073/pnas.0502848102

[47]   S. Kumar, Y. Lei, N.H. Alshareef, M.A.Q. Lopez, K.N. Salama, Biofunctionalized two-dimensional Ti_3C_2 MXenes for ultrasensitive detection of cancer biomarker, Biosens. Bioelectron. 121 (2018) 243 – 249. https://doi.org/10.1016/j.bios.2018.08.076

[48]   D. Wu, M. Wu, J. Yang, H. Zhang, K. Xie, C. Lin, A. Yu, J. Yu, L. Fu, Delaminated $Ti_3C_2T_x$ (MXene) for electrochemical carbendazim sensing, Mater. Lett. 236 (2019) 412 – 415. https://doi.org/10.1016/j.matlet.2018.10.150

[49]   Q. Shan, X. Mu, M. Alhabeb, C.E. Shuck, D. Pang, X. Zhao, X. Chu, Y. Wei, F. Du, G. Chen, Y. Gogotsi, Y. Gao, Y. Dall'Agnese, Two-dimensional vanadium carbide (V_2C) MXene as electrode for supercapacitors with aqueous electrolytes, Electrochem. Commun. 96 (2018) 103 – 107. https://doi.org/10.1016/j.elecom.2018.10.012

[50] W. Zhi, S. Xiang, R. Bian, R. Lin, K. Wu, T. Wang, D. Cai, Study of MXene-filled polyurethane nanocomposites prepared via an emulsion method, Compos. Sci. Technol. 168 (2018) 404 – 411. https://doi.org/10.1016/j.compscitech.2018.10.026

[51] L. Yang, W. Zheng, P. Zhang, J. Chen, W.B. Tian, Y.M. Zhang, Z.M. Sun, MXene/CNTs films prepared by electrophoretic deposition for supercapacitor electrodes, J. Electroanal. Chem. 830 – 831 (2018) 1 – 6. https://doi.org/10.1016/j.jelechem.2018.10.024

[52] A.R. Wojciechowska, T. Wojciechowski, W. Ziemkowska, L. Chlubny, A. Olszyna, A.M. Jastrzebska, Surface interactions between 2D Ti_3C_2/Ti_2C MXenes and lysozyme, Appl. Surf. Sci. 473 (2019) 409 – 418. https://doi.org/10.1016/j.apsusc.2018.12.081

[53] X. Zou, G. Li, Q. Wang, D. Tang, B. Wu, X. Wang, Energy storage properties of selectively functionalized Cr-group MXenes, Comput. Mater. Sci. 150 (2018) 236 – 243. https://doi.org/10.1016/j.commatsci.2018.04.014

[54] M. Magnuson, J. Halim, L. Naslund, Chemical bonding in carbide MXene nanosheets, J. Electron. Spectrosc. Relat. Phenom. 224 (2018) 27 – 32. https://doi.org/10.1016/j.elspec.2017.09.006

[55] C. Eames, M.S. Islam, Ion intercalation into two-dimensional transition-metal carbides: global screening for new high-capacity battery materials, J. Am. Chem. Soc. 136 (2014) 16270 – 16276. https://doi.org/10.1021/ja508154e

[56] G.R. Berdiyorov, K.A. Mahmoud, Effect of surface termination on ion intercalation selectivity of bilayer $Ti_3C_2T_2$ (T = F, O and OH) MXene, Appl. Surf. Sci. 416 (2017) 725 – 730. https://doi.org/10.1016/j.apsusc.2017.04.195

[57] A.N. Enyashin, A.L. Ivanovskii, Structural, electronic properties and stability of MXenes Ti_2C and Ti_3C_2 functionalized by methoxy groups, J. Phys. Chem. 117 (2013) 13637 – 13643. https://doi.org/10.1021/jp401820b

[58] J. Zhu, U. Schwingenschlogl, P and Si functionalized MXenes for metal-ion battery applications, 2D Mater. 4 (2017) 025073. https://doi.org/10.1088/2053-1583/aa69fe

[59] X. Gao, Y. Zhou, Y. Tan, Z. Cheng, B. Yang, Y. Ma, Z. Shen, J. Jia, Exploring adsorption behavior and oxidation mechanism of mercury on monolayer Ti_2CO_2 (MXenes) from first principles, Appl. Surf. Sci. 464 (2019) 53 – 60. https://doi.org/10.1016/j.apsusc.2018.09.071

[60]　I.R. Shein, A.L. Ivanovskii, Graphene-like nanocarbides and nanonitrides of *d* metals (MXenes): synthesis, properties and simulation, Micro Nano Lett. 8 (2013) 59 – 62. https://doi.org/10.1049/mnl.2012.0797

[61]　J. Guo, Y. Sun, B. Liu, Q. Zhang, Q. Peng, Two-dimensional scandium-based carbides (MXene): band gap modulation and optical properties, J. Alloys Compd. 712 (2017) 752 – 759. https://doi.org/10.1016/j.jallcom.2017.04.149

[62]　M. Khazaei, A. Ranjbar, M. Ghorbani-Asl, M. Arai, T. Sasaki, Y. Liang, S. Yunoki, Nearly free electron states in MXenes, Phys. Rev. B 93 (2016) 205125-1 – 205125-10. https://doi.org/10.1103/physrevb.93.205125

[63]　M. Khazaei, M. Arai, T. Sasaki, C. Chung, N.S. Venkataramanan, M. Estili, Y. Sakka, Y. Kawazoe, Novel electronic and magnetic properties of two-dimensional transition metal carbides and nitrides, Adv. Funct. Mater. 17 (2013) 2185 – 2192. https://doi.org/10.1002/adfm.201202502

[64]　M. Khazaei, M. Arai, T. Sasaki, M. Estili, Y. Sakka, Two-dimensional molybdenum carbides: potential thermoelectric materials of the MXene family, Phys. Chem. Chem. Phys. 16 (2014) 7841 – 7849. https://doi.org/10.1039/c4cp00467a

[65]　M. Han, X. Yin, H. Wu, Z. Hou, C. Song, X. Li, L. Zhang, L. Cheng, Ti_3C_2 MXenes with modified surface for high-performance electromagnetic absorption and shielding in the X-band, ACS Appl. Mater. Interfaces 8 (2016) 21011 – 21019. https://doi.org/10.1021/acsami.6b06455

[66]　A. Shahzad, K. Rasool, W. Miran, M. Nawaz, J. Jang, K.A. Mahmoud, D.S. Lee, Mercuric ion capturing by recoverable titanium carbide magnetic nanocomposite, J. Hazard. Mater. 344 (2018) 811 – 818. https://doi.org/10.1016/j.jhazmat.2017.11.026

[67]　M. Cao, Y. Cai, P. He, J. Shu, W. Cao, J. Yuan, 2D MXenes: electromagnetic property for microwave absorption and electromagnetic interference shielding, Chem. Eng. J. 359 (2019) 1265 – 1302. https://doi.org/10.1016/j.cej.2018.11.051

[68]　H. Yang, J. Dai, X. Liu, Y. Lin, J. Wang, L. Wang, F. Wang, Layered $PVB/Ba_3Co_2Fe_{24}O_{41}/Ti_3C_2$ MXene composite: enhanced electromagnetic wave absorption properties with high impedance match in a wide frequency range, Mater. Chem. Phys. 200 (2017) 179 – 186. https://doi.org/10.1016/j.matchemphys.2017.05.057

[69]　X. Su, J. Zhang, H. Mu, J. Zhao, Z. Wang, Z. Zhao, C. Han, Z. Ye, Effects of etching temperature and ball milling on the preparation and capacitance of Ti_3C_2

MXene, J. Alloys Compd. 752 (2018) 32 – 39.
https://doi.org/10.1016/j.jallcom.2018.04.152

[70]   R. Meshkian, Q. Tao, M. Dahlqvist, J. Lu, L. Hultman, J. Rosen, Theoretical stability and materials synthesis of a chemically ordered MAX phase, Mo_2ScAlC_2, and its two-dimensional derivate Mo_2ScC_2 MXene, Acta Mater. 125 (2017) 476 – 480. https://doi.org/10.1016/j.actamat.2016.12.008

[71]   S. Zhao, X. Meng, K. Zhu, F. Du, G. Chen, Y. Wei, Y. Gogotsi, Y. Gao, Li-ion uptake and increase in interlayer spacing of Nb_4C_3 MXene, Energy Storage Mater. 8 (2017) 42 – 48. https://doi.org/10.1016/j.ensm.2017.03.012

[72]   J. Zhou, S. Gao, Z. Guo, Z. Sun, Ti-enhanced exfoliation of V_2AlC into V_2C MXene for lithium-ion battery anodes, Ceram. Int. 43 (2017) 11450 – 11454. https://doi.org/10.1016/j.ceramint.2017.06.016

[73]   W. Feng, H. Luo, Y. Wang, S. Zeng, Y. Tan, H. Zhang, S. Peng, Ultrasonic assisted etching and delaminating of Ti_3C_2 Mxene, Ceram. Int. 44 (2018) 7084 – 7087. https://doi.org/10.1016/j.ceramint.2018.01.147

[74]   C. Peng, P. Wei, X. Chen, Y. Zhang, F. Zhu, Y. Cao, H. Wang, H. Yu, F. Peng, A hydrothermal etching route to synthesis of 2D MXene (Ti_3C_2, Nb_2C): enhanced exfoliation and improved adsorption performance, Ceram. Int. 44 (2018) 18886 – 18893. https://doi.org/10.1016/j.ceramint.2018.07.124

[75]   M. Ghidiu, M.R. Lukatskaya, M. Zhao, Y. Gogotsi, M.W. Barsoum, Conductive two-dimensional titanium carbide 'clay' with high volumetric capacitance, Nature 516 (2014) 78 – 82. https://doi.org/10.1038/nature13970

[76]   F. Liu, A. Zhou, J. Chen, J. Jia, W. Zhou, L. Wang, Q. Hu, Preparation of Ti_3C_2 and Ti_2C MXenes by fluoride salts etching and methane adsorptive properties, Appl. Surf. Sci. 416 (2017) 781 – 789. https://doi.org/10.1016/j.apsusc.2017.04.239

[77]   Z. Xu, G. Liu, H. Ye, W. Jin, Z. Cui, Two-dimensional MXene incorporated chitosan mixed-matrix membranes for efficient solvent dehydration, J. Membr. Sci. 563 (2018) 625 – 632. https://doi.org/10.1016/j.memsci.2018.05.044

[78] L. Wang, H. Zhang, B. Wang, C. Shen, C. Zhang, Q. Hu, A. Zhou, B. Liu, Synthesis and electrochemical performance of $Ti_3C_2T_x$ with hydrothermal process, Electron. Mater. Lett. 12 (2016) 702 – 710. https://doi.org/10.1007/s13391-016-6088-z

[79] X. Xie, Y. Xue, L. Li, S. Chen, Y. Nie, W. Ding, Z. Wei, Surface Al leached Ti_3AlC_2 as a substitute for carbon for use as a catalyst support in a harsh corrosive

electrochemical system, Nanoscale 6 (2014) 11035 – 11040.
https://doi.org/10.1039/c4nr02080d

[80]   M. Naguib, R.R. Unocic, B.L. Armstrong, J. Nanda, Large-scale delamination of
multi-layers transition metal carbides and carbonitrides "MXenes", Dalton Trans. 44
(2015) 9353 – 9358. https://doi.org/10.1039/c5dt01247c

[81]   K. Singh, S. Arora, Removal of synthetic textile dyes from wastewaters: A critical
review on present treatment technologies, Crit. Rev. Environ. Sci. Technol. 41 (2011)
37–41.

[82]   T. Ngulube, J. Ray, V. Masindi, A. Maity, An update on synthetic dyes adsorption
onto clay based minerals : A state-of-art review, J. Environ. Manage. 191 (2017) 35–
57. https://doi.org/10.1016/j.jenvman.2016.12.031

[83]   A.K. Verma, R.R. Dash, P. Bhunia, A review on chemical coagulation/flocculation
technologies for removal of colour from textile wastewaters, J. Environ. Manage. 93
(2012) 154–168. https://doi.org/10.1016/j.jenvman.2011.09.012

[84]   S. Chowdhury, R. Balasubramanian, Recent advances in the use of graphene-
family nanoadsorbents for removal of toxic pollutants from wastewater, Adv. Colloid
Interface Sci. 204 (2014) 35–56. https://doi.org/10.1016/j.cis.2013.12.005

[85]   L. Wu, X. Lu, Z. Wu, Y. Dong, X. Wang, S. Zheng, J. Chen, 2D transition metal
carbide MXene as a robust biosensing platform for enzyme immobilization and
ultrasensitive detection of phenol, Biosens. Bioelectron. 107 (2018) 69–75
https://doi.org/10.1016/j.bios.2018.02.021.

[86]   P. Gu, S. Zhang, Ch. Zhang, X. Wang, A. Khan, T. Wen, B. Hu, A. Alsaedi, T.
Hayat, Two-dimensional MAX-derived titanate nanostructures for efficient removal of
Pb(II), Dalt. Trans. 48 (2019) 2100–2107. https://doi.org/10.1039/c8dt04301a

[87]   Y. Ying, Y. Liu, X. Wang, Y. Mao, W. Cao, P. Hu, X. Peng, Two-dimensional
titanium carbide for efficiently reductive removal of highly toxic chromium (VI) from
water, ACS Appl. Mater. Interfaces. 7 (2015) 1795–1803.
https://doi.org/10.1021/am5074722

[88]   X. Zhu, B. Liu, H. Hou, Z. Huang, K. Mohammed, L. Huang, X. Yuan, D. Guo, J.
Hu, J. Yang, Alkaline intercalation of Ti_3C_2 MXene for simultaneous electrochemical
detection of Cd (II), Pb (II), Cu (II) and Hg (II), Electrochim. Acta. 248 (2017) 46–57.
https://doi.org/10.1016/j.electacta.2017.07.084

[89]   Z. Wei, Z. Peigen, T. Wubian, Q. Xia, Z. Yamei, Alkali treated $Ti_3C_2T_x$ MXenes and their dye adsorption performance, Mater. Chem. Phys. 206 (2018) 270–276. https://doi.org/10.1016/j.matchemphys.2017.12.034

[90]   F. Meng, M. Seredych, C. Chen, V. Gura, S. Mikhalovsky, R. Susan, MXene sorbents for removal of urea from dialysate–a step towards the wearable artificial kidney, ACS Nano. 12 (2018) 10518–10528. https://doi.org/10.1021/acsnano.8b06494

[91]   Y. Zhang, Z. Zhou, J. Lan, P. Zhang, Prediction of $Ti_3C_2O_2$ MXene as an effective capturer of formaldehyde, Appl. Surf. Sci. 469 (2019) 770–774. https://doi.org/10.1016/j.apsusc.2018.11.018

[92]   J. Guo, H. Fu, G. Zou, Q. Zhang, Z. Zhang, Theoretical interpretation on lead adsorption behavior of new two-dimensional transition metal carbides and nitrides, J. Alloys Compd. 684 (2016) 504–509. https://doi.org/10.1016/j.jallcom.2016.05.217

[93]   A. Shahzad, K. Rasool, W. Miran, M. Nawaz, J. Jang, K. A. Mahmoud, D. Sung Lee, Two-dimensional $Ti_3C_2T_x$ MXene nanosheets for efficient copper removal from water, ACS Sustain. Chem. Eng. 5 (2017) 11481–11488. https://doi.org/10.1021/acssuschemeng.7b02695

[94]   A. Kayvani-Fard, G. Mckay, R. Chamoun, T. Rhadfi, H. Preud Homme, M.A. Atieh, Barium removal from synthetic natural and produced water using MXene as two dimensional (2-D) nanosheet adsorbent, Chem. Eng. J. 317 (2017) 331–342. https://doi.org/10.1016/j.cej.2017.02.090

[95]   J. Yang, S.-Z. Zhang, T.-L. Ji, S.-H. Wei, Adsorption activities of O, OH, F and Au on two-dimensional Ti_2C and Ti_3C_2 surfaces, Acta Physico-Chimica Sin. 31 (2015) 369–376.

[96]   M.A. Iqbal, S.I. Ali, A. Tariq, M.Z. Iqbal, S. Rizwan, Improved organic dye degradation using highly efficient MXene composites, Preprints. 1 (2018) 1–12. https://doi.org/10.20944/preprints201811.0386.v1

[97]   Q. Zhang, J. Tengb, G. Zoua, Q. Peng, Q. Dub, T. Jiao, J. Xianga, Efficient phosphate sequestration for water purification by unique sandwichlike MXene/magnetic iron oxide nanocomposites, Nanoscale. 8 (2016) 7085–7093. https://doi.org/10.1039/c5nr09303a

[98]   W. Mu, S. Du, Q. Yu, X. Li, H. Wei, Y. Yang, Improving barium ions adsorption on two-dimensional titanium carbide by surface modification, Dalt. Trans. 47 (2018) 8375–8381. https://doi.org/10.1039/c8dt00917a

[99]   L. Wang, L. Yuan, K. Chen, Y. Zhang, Q. Deng, S. Du, Q. Huang, L. Zheng, J. Zhang, Z. Chai, M.W. Barsoum, X. Wang, W. Shi, Loading actinides in multilayered structures for nuclear waste treatment: the first case study of uranium capture with vanadium carbide MXene, ACS Appl. Mater. Interfaces. 8 (2016) 16396–16403. https://doi.org/10.1021/acsami.6b02989

[100] R.P. Pandey, K. Rasool, P.A. Rasheed, K.A. Mahmoud, Reductive sequestration of toxic bromate from drinking water using lamellar 2D $Ti_3C_2T_X$ (MXene), ACS Sustain. Chem. Eng. 6 (2018) 7910–7917. https://doi.org/10.1021/acssuschemeng.8b01147

[101] X. Yu, Y. Li, J. Cheng, Z. Liu, Q. Li, W. Li, X. Yang, B. Xiao, Monolayer Ti_2CO_2: A promising candidate for NH_3 sensor or capturer with high sensitivity and selectivity, ACS Appl. Mater. Interfaces. 7 (2015) 13707–13713. https://doi.org/10.1021/acsami.5b03737

[102] A. Morales-Garcia, A. Fernandez-Fernandez, F. Vines, F. Illas, CO_2 abatement using two-dimensional MXene carbides, J. Mater. Chem. A. (2018) 3381–3385. https://doi.org/10.1039/c7ta11379j

[103] S. Ma, D. Yuan, Z. Jiao, T. Wang, X. Dai, Monolayer Sc_2CO_2: A promising candidate as SO_2 gas sensor or capturer, J. Phys. Chem. C. 121 (2017) 24077–24084. https://doi.org/10.1021/acs.jpcc.7b07921

[104] Y. Zhang, Z. Zhou, J. Lan, C. Ge, Z. Chai, Theoretical insights into the uranyl adsorption behavior on vanadium carbide MXene, Appl. Surf. Sci. 426 (2017) 572–578. https://doi.org/10.1016/j.apsusc.2017.07.227

[105] Y. Zhang, J. Lan, L. Wang, Q. Wu, C. Wang, T. Bo, Z. Chai, W. Shi, Adsorption of uranyl species on hydroxylated titanium carbide nanosheet : A first-principles study, J. Hazard. Mater. 308 (2016) 402–410. https://doi.org/10.1016/j.jhazmat.2016.01.053

[106] L. Wang, W. Tao, L. Yuan, Z. Liu, Q. Huang, Z. Chai, J.K. Gibson, W. Shi, Rational control of interlayer space inside two-dimensional titanium carbides for highly efficient uranium removal and imprisoning, ChemComm. 53 (2017) 12084–12087. https://doi.org/10.1039/c7cc06740b

[107] Q. Tang, Z. Zhou, P. Shen, Are MXenes promising anode materials for Li ion batteries? computational studies on electronic properties and Li storage capability of Ti_3C_2 and $Ti_3 C_2X_2$ (X= F, OH) monolayer, J. Am. Chem. Soc. 134 (2012) 16909–16916. https://doi.org/10.1021/ja308463r

[108] S. Zhao, W. Kang, J. Xue, Role of strain and concentration on the Li adsorption and diffusion properties on Ti$_2$C layer, J. Phys. Chem. C. 118 (2014) 14983–14990. https://doi.org/10.1021/jp504493a

[109] S.N. Ahmed, W. Haider, Heterogeneous photocatalysis and its potential applications in water and wastewater treatment: a review. Nanotechnol. 29 (2018) p.342001. https://doi.org/10.1088/1361-6528/aac6ea

[110] C.B. Ong, L.Y. Ng, A.W. Mohammed, A review of ZnO nanoparticles as solar photocatalysis: synthesis, mechanisms and applications, Renew. Sust. Energ. Rev. 81 (2018) 536–551.

[111] C. Zhang, V. Nicolosi, Graphene and MXene-based transparent conductive electrodes and supercapacitors. Energy Storage Mater. 16 (2019) 102-125. https://doi.org/10.1016/j.ensm.2018.05.003

[112] S. Kumar, Y. Lei, N.H. Alshareef, M.A. Quevedo-Lopez, K.N. Salama, Biofunctionalized two-dimensional Ti$_3$C$_2$ MXenes for ultrasensitive detection of cancer biomarker. Biosens. Bioelectron. 121 (2018) 243-249. https://doi.org/10.1016/j.bios.2018.08.076

[113] S. Sun, C. Liao, A.M. Hafez, H. Zhu, S. Wu, Two-dimensional MXenes for energy storage. Chem. Eng. J. 338 (2018) 27-45. https://doi.org/10.1016/j.cej.2017.12.155

[114] X. Bai, C. Ling, L. Shi, Y. Ouyang, Q. Li, J. Wang, Insight into the catalytic activity of MXenes for hydrogen evolution reaction. Sci. Bull. 63 (2018) 1397-1403. https://doi.org/10.1016/j.scib.2018.10.006

[115] H. Zhang, M. Li, J. Cao, Q. Tang, P. Kang, C. Zhu, M. Ma, 2D A-Fe$_2$O$_3$ doped Ti$_3$C$_2$ MXene composite with enhanced visible light photocatalytic activity for degradation of Rhodamine B. Ceram. Int. 44 (2018) 19958-19962. https://doi.org/10.1016/j.ceramint.2018.07.262

[116] Y. Gao, L. Wang, A. Zhou, Z. Li, J. Chen, H. Bala, Q. Hu, X. Cao, Hydrothermal synthesis of TiO$_2$/Ti$_3$C$_2$ nanocomposites with enhanced photocatalytic activity. Mater. Lett. 150 (2015) 62-64. https://doi.org/10.1016/j.matlet.2015.02.135

[117] Y. Lu, M. Yao, A. Zhou, Q. Hu, L. Wang, Preparation and photocatalytic performance of Ti$_3$C$_2$/TiO$_2$/CuO ternary nanocomposites. J. Nanomater. 2017 (2017) 1-5. https://doi.org/10.1155/2017/1978764

[118] W. Hu, C. Peng, W. Luo, M. Lv, X. Li, D. Li, Q. Huang, C. Fan, Graphene-based antibacterial paper. ACS Nano. 4 (2010) 4317-4323. https://doi.org/10.1021/nn101097v

[119] S. Liu, T.H. Zeng, M. Hofmann, E. Burcombe, J. Wei, R. Jiang, J. Kong, Y. Chen, Antibacterial activity of graphite, graphite oxidem graphene oxide, and reduced graphene oxide: membrane and oxidative stress. ACS Nano. 5 (2011) 6971-6980. https://doi.org/10.1021/nn202451x

[120] X. Yang, J. Li, T. Liang, C. Ma, Y. Zhang, H. Chen, N. Hanagata, H. Su, M. Xu, Antibacterial activity of two-dimensional MoS$_2$ sheets. Nanoscale 6 (2014) 10126-10133. https://doi.org/10.1039/c4nr01965b

[121] F. Alimohammadi, M. Sharifian Gh, N.H. Attanayake, A.C. Thenuwara, Y. Gogotsi, B. Anasori, D.R. Strongin, Antimicrobial Properties of 2D MnO$_2$ and MoS$_2$ Nanomaterials Vertically Aligned on Graphene Materials and Ti$_3$C$_2$ MXene, Langmuir 34 (2018) 7192-7200. https://doi.org/10.1021/acs.langmuir.8b00262

[122] B.M. Jun, S. Kim, J. Heo, C.M. Park, N. Her, M. Jang, Y. Huang, J. Han, Y. Yoon, Review of MXenes as new nanomaterials for energy storage/delivery and selected environmental applications. Nano Res. (2018) 1-17. https://doi.org/10.1007/s12274-018-2225-3

[123] K. Rasool, M. Helal, A. Ali, C.E. Ren, Y. Gogotsi, K.A. Mahmoud, Antibacterial activity of Ti$_3$C$_2$T$_x$ MXene. ACS Nano. 10 (2016) 3674-3684. https://doi.org/10.1021/acsnano.6b00181

[124] A. Arabi Shamsabadi, M. Sharifian Gh, B. Anasori, M. Soroush,Antimicrobial mode-of-action of colloidal Ti$_3$C$_2$T$_x$ MXene nanosheets. ACS Sustain. Chem. Eng. 6(2018)16586-16596. https://doi.org/10.1021/acssuschemeng.8b03823

[125] E.A. Mayerberger, R.M. Street, R.M. McDaniel, M.W. Barsoum, C.L. Schauer, Antibacterial properties of electrospun Ti$_3$C$_2$T$_z$ (MXene)/chitosan nanofibers. RSC Adv. 8(2018)35386-35394. https://doi.org/10.1039/c8ra06274a

[126] H. Lin, Y. Chen, J. Shi, Insights into 2D MXenes for versatile biomedical applications: current advances and challenges ahead. Adv. Sci. 5(2018), p.1800518. https://doi.org/10.1002/advs.201800518

[127] B. Wu, Membrane-based technology in greywater reclamation: a review. Sci. Total. Environ. 656 (2019) 184-200. https://doi.org/10.1016/j.scitotenv.2018.11.347

[128] E. Yang, K.J. Chae, M.J. Choi, Z. He, I.S. Kim, Critical review of bioelectrochemical systems integrated with membrane-based technologies for desalination, energy self-sufficiency, and high-efficiency water and wastewater treatment. Desalination 452 (2019) 40-67. https://doi.org/10.1016/j.desal.2018.11.007

[129] K.H. Chu, Y. Huang, M. Yu, J. Heo, J.R.V. Flora, A. Jang, M. Jang, C. Jung, C.M. Park, D.H. Kim, Y. Yoon, Evaluation of graphene oxidecoated ultrafiltration membranes for humid acid removal at different pH and conductivity conditions. Sep. Purif. Technol. 181 (2017) 139-147. https://doi.org/10.1016/j.seppur.2017.03.026

[130] J. Ma, X. Guo, Y. Ying, D. Liu, C. Zhong, Composite ultrafiltration membrane tailored by MOF@GO with highly improved water purification performance. Chem. Eng. J. 313 (2017) 890-898. https://doi.org/10.1016/j.cej.2016.10.127

[131] G. Liu, J. Shen, Q. Liu, G. Liu, J. Xiong, J. Yang, W. Jin, Ultrathin two-dimensional MXene membrane for pervaporation desalination. J. Membrane Sci. 548 (2018) 548-558. https://doi.org/10.1016/j.memsci.2017.11.065

[132] R. Han, Y. Xie, X. Ma, Crosslinked P84 copolyimide/MXene mixed matrix membrane with excellent solvent resistance and permselectivity. Chin. J. Chem. Eng. (2018). https://doi.org/10.1016/j.cjche.2018.10.005

[133] R.P. Pandey, K. Rasool, V.E. Madhavan, B. Aissa, Y. Gogotsi, K.A. Mahmoud, Ultrahigh-flux and fouling-resistant membranes based on layered silver/MXene $(Ti_3C_2T_x)$ nanosheets. J. Mater. Chem A. 6 (2018) 3522-3533. https://doi.org/10.1039/c7ta10888e

MXenes: Fundamentals and Applications   Materials Research Forum LLC
Materials Research Foundations **51** (2019) 61-73   doi: https://doi.org/10.21741/9781644900253-3

Chapter 3

Two-Dimensional MXene as a Promising Material for Hydrogen Storage

Jin-Chung Sin[1*], Jian-Ai Quek[1], Pei-Sian Ng[1], Sze-Mun Lam[2]

[1] Department of Petrochemical Engineering, Faculty of Engineering and Green Technology, Universiti Tunku Abdul Rahman, Jalan Universiti, Bandar Barat, 31900 Kampar, Perak, Malaysia

[2] Department of Environmental Engineering, Faculty of Engineering and Green Technology, Universiti Tunku Abdul Rahman, Jalan Universiti, Bandar Barat, 31900 Kampar, Perak, Malaysia

sinjc@utar.edu.my*

Abstract

Hydrogen is generally the cleanest, a sustainable and renewable energy source with a considerably reduced impact on the environment. The adoption of H_2 energy has been expected to replace the depleting carbon-based fuels gradually. An appropriate hydrogen storage system which capability of charging and discharging large amounts of hydrogen with fast enough kinetics is required to meet the broad on-board applications. Transition metal carbides and nitrides (MXenes) are a new family of two-dimensional inorganic hybrid material with ultra-large interlayer spacing, excellent biocompatibility, environmental benignity, surface hydrophilic property, and good electric conductivity. Their prominent features of MXenes are attractive and ideal for hydrogen storage. This chapter provides a review dealing with the preparation and use of MXenes used in hydrogen storage as well as to reveal computational and theoretical studies on hydrogen storage.

Keywords

MXene, Hydrogen Storage, Computational, Theoretical, Two-Dimensional

Contents

1.   Introduction

Economic growth, environmental pollution, and energy shortage are causing serious challenges. Therefore, renewable and environmentally friendly energy sources are highly demanded to fulfil future energy requirements. H_2 is considered as one of the most promising green energy carriers due to its high energy density, abundance and non-polluting nature [1]. Over the past few decades, an effective and reliable H_2 storage method has been widely experimented by many researchers [2,3]. Numerous researches have been focused on improving the storage capacity of H_2 and H_2 adsorption/desorption behavior. Of all the factors been analyzed, developing of new-concept materials with extraordinary performance is of significance and urgency.

Two-dimensional (2-D) materials [4-7] have garnered enormous scientific interest because of their outstanding physicochemical properties in comparison with their bulk materials. One of the popular 2-D materials, graphene has been considered as an ideal candidate for its great potential in a wide range of applications [4,7]. In recent years, 2-D transition metal carbides, carbonitrides and nitrides (denoted as MXenes) have developed promptly since the discovery of Ti_3C_2 in 2011 [8]. These layered materials paved its way into various potential applications based on a similar structure as graphene, which exhibited widespread opportunities in next generation energy devices. The members of MXene family have been heavily investigated owing to their distinctive characteristics like ultra-large interlayer spacing, excellent biocompatibility, environmentally friendly and obvious security capability [9,10]. In addition to the 2-D lamellar structure, MXenes also possessed hydrophilic nature and superb electrical conductivity [11].

MXenes: Fundamentals and Applications   Materials Research Forum LLC
Materials Research Foundations **51** (2019) 61-73   doi: https://doi.org/10.21741/9781644900253-3

Moreover, they are capable of hosting a wide variety of cations between their layers [12]. Above outstanding characteristics have resulted in MXenes to be promising energy storage materials. Herein, this chapter provides a review dealing with the preparation and use of MXenes used in hydrogen storage as well as to reveal computational and theoretical study on hydrogen storage.

2.   Family of Mxenes

Mxenes are a new family of 2-D transition metal carbides, carbonitrides and nitrides. MXenes are made from a bulk crystal called MAX with the general formula $M_{n+1}AX_n$, where n = 1–3, M represents a transition metal, A denotes an element for examples aluminum or silicon, and X stands for carbon or nitrogen (Fig. 1) [13]. In an etching process with chemicals, the A layer removed from MAX phase and resulted in monolayer, with surfaces functionalized, T_x, by functional groups of oxide (–O), hydroxyl (–OH) or fluorine (–F) to provide the general formula $M_{n+1}X_nT_x$ [13]. According to Lei et al. [14], F terminations were unstable and can be substituted by –OH groups when rinsing with water. Therefore, OH⁻ and/or O⁻ terminated Mxenes were more stable. Also, -OH groups able to convert into O terminations by heating at high temperature or adsorbing metal ion. The O-terminated MXene can also decompose into naked MXene when interacted with Mg, Al and Ca metals.

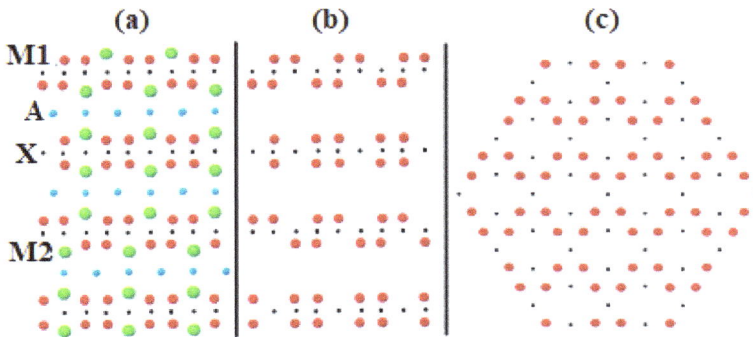

Figure 1 (a) Cross observation of hypothetically projected MAX phase; (M1, M2) 2AX and the (b) resultant material after etching of M2 and A elements. (c) Top observation of 2-D sheet from (b), demonstrating well-organized vacancies [13].

MXenes: Fundamentals and Applications   Materials Research Forum LLC
Materials Research Foundations **51** (2019) 61-73   doi: https://doi.org/10.21741/9781644900253-3

3.   Structural properties of Mxenes

Full geometry optimization for MXenes was determined as the first step [15]. Modeling is important to study MXene's structure and determine the structure properties. Modeling helped to predict the positions and orientations of surface groups. Mxenes needed to model with different surface groups so that to more precisely reveal the complicated structure of the material system. Furthermore, hydrogen bonding and van der Waals bonding in the interactions between the MXene layers needed to be studied because most of the Mxene is multilayer stacking.

According to Chen et al. [16], structural properties such as crystallinity and phase structure of Mxenes were investigated using X-ray Diffraction (XRD) and differential scanning calorimetry (DSC). The morphology, size, and shape of MXene scan be identified using scanning electron microscopy (SEM) and transmission electron microscopy (TEM). The elemental composition of MXenes was analyzed using energy dispersive X-ray spectroscopy (EDX). The binding energies were investigated using X-ray photon spectroscopy (XPS), and de-hydrogenation properties were examined by volumetric release (VR).

4.   Preparation of Mxenes

MXenes are mostly produced by selectively etching weakly bonded A layers from the MAX phases using hydrofluoric acid (HF), ammonium bifluoride (NH_4HF_2) and a mixture of hydrochloric acid (HCl) and lithium fluoride (LiF) solutions [17,18]. In a study by Naguib et al. [8], Al layers were extracted from Ti_3AlC_2 phase to form $Ti_3C_2T_x$. The chemical reactions of Ti_3AlC_2 with HF solutions are reported as follows:

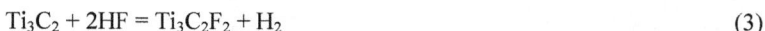

$$Ti_3AlC_2 + 3HF = AlF_3 + 3/2H_2 + Ti_3C_2 \tag{1}$$

$$Ti_3C_2 + 2H_2O = Ti_3C_2(OH)_2 + H_2 \tag{2}$$

$$Ti_3C_2 + 2HF = Ti_3C_2F_2 + H_2 \tag{3}$$

Formation of -OH and-F terminations can be observed in Reactions (2) and (3). The solids were centrifuged and washed with H_2O. In the absence of delamination, multilayered structures of Mxenes were found. The sample was stirred with DMSO (dimethyl sulfoxide) for 18 h, and the separated intercalated powder was reported to bemono- or few-layer Mxenes [19]

This technique is been utilized to synthesize almost all types of MXene from Al-containing MAX phases. The conditions of chemical etching process (HF concentration and time) depended on the temperature and particle size of MAX. [14]. For instance, decreasing the size of the MAX phase via ball milling or attrition efficiently reduced the etching time and/or HF concentration [20,21]. Moreover, variances in M–Al binding energies for MAX phase of different M elements needed diverse etching conditions. For instance, the smaller Nb-Al bond energy in Nb_2AlC led to shortening etching time and decreased HF concentration [21,22]. Therefore, proper etching conditions are significant influences in obtaining high yield and permitting the complete conversion to Mxene.

In another study, Halim et al. [23] used the ammonium bifluoride (NH_4HF_2) as a milder etchant instead of the perilous HF. Its use also led to concomitant intercalation of cations during the etching process. Therefore, this strategy is more appropriate in the preparation of delaminated Mxenes. Since the intercalation and etching reactions carried out concurrently, the chemical processes can be given as:

$$Ti_3AlC_2 + 3NH_4HF_2 = (NH_4)_3AlF_6 + 3/2H_2 + Ti_3C_2 \qquad (4)$$

$$Ti_3C_2 + aNH_4HF_2 + bH_2O = (NH_3)_c(NH_4)_dTi_3C_2(OH)_xF_y \qquad (5)$$

The atomic layers in $Ti_3C_2T_x$ are more congruously spaced and appeared to be attached together owing to the slower and less vigorous reactions as well as the intercalation of both NH_3 and NH_4^+.

5. Mxenes for hydrogen storage

Hu et al. [24] stated that H_2 was able to be stored in solid materials, liquid phases or gaseous phases. There are two categories of solid-state storage materials: chemisorption of dissociated hydrogen atoms and physisorption of intact hydrogen molecules. In the presence of strong covalent bond, the H-atoms bound with metal hydrides with binding energy more than 0.4 eV was recognized as chemisorption. The chemisorption occurred at high temperature to overcome or penetrate the activation energy barrier to produce chemical bond, which generally dissociated hydrogen molecules into two hydrogen atoms. By binding energy, less than 0.1 eV, the hydrogen molecule bound with metal hydrides via a weak van der Waals force can be known as physisorption. On the other hand, the binding energy of H_2 that is located in the range of 0.1 to 0.4 eV is known as Kubas interaction. Furthermore, the Kubas interaction can be determined by these features, namely (i) lengthen of H-H bond in the absence of rupture, (ii) electron migration to vacant d orbital of metal from filled H_2 molecule bonding orbital and (iii) back donation to the H_2 anti-bonding orbitals from filled metal d orbital [25].

However, chemisorption and physisorption exhibited their own limitations. The chemisorption process restricted with the challenging of hydrogen release at moderate temperature owing to strong bonding between H atoms and metal hydrides or complex chemical hydrides (40−80 kJ/mol) [26,27]. The physisorption was known as suitable sorbent material only when exhibited both these properties: (i) appropriate binding energy ($\sim$0.2–0.3 eV) with H_2molecules and (ii) large specific surface area [26,28]. Thus, the high surface area of sorption-based storage materials was more favorable.

Bulk crystal of Mxenes provided a large surface area per volume thus helping in energy storage applications. 2-D crystals like Mxenes which derived by exfoliating MAX phases also assisted in hydrogen storage. The members of MXenes family such as Ti_2C, Sc_2C, and V_2C phases can bind the H_2 molecules with Kubas interaction have been reported by many researchers as capable reversible hydrogen storage material [24,29]. For example, the hydrogen storage behaviour on two-dimensional Sc_2C has been systematically investigated by first-principles total energy pseudopotential calculations. Their investigations showed that the hydrogen storage capacity for all the adsorbed hydrogen atoms and molecules was 9.0 wt%, which met the gravimetric storage capacity target (5.5 wt.%) set by United States Department of Energy. They also further reported that the binding energy of 0.164 eV for the H_2 molecule by Kubas-type interaction was appropriate as a reversible hydrogen storage material at ambient conditions [29]. Nevertheless, there are still many unexplored MXenes which can serve as excellent materials towards hydrogen storage applications.

6.   Computational and theoretical study on hydrogen storage over MXenes

Computation method for hydrogen storage calculation was reported by numerous theoretical studies using MXenes. Among numerous computation methods, the density functional theory (DFT) was chosen as a reliable way to predict various atomic scaled physical and chemical states of the MXenes. The popular DFT method had implemented CASTEP code for all first-principles total energy pseudopotential calculation execution of hydrogen storage potential in MXenes materials [25]. However, the DFT method may fail to predict band gap value, van der Waals interaction and strong correlated materials properties of the MXenes products [30]. To overcome these deficiencies, DFT can be combined with other coupled computational methods to improve the characterization of MXenes properties.

Local density approximation (LDA) was employed among other coupled computational methods to understand the exchange-correlation functional of MXenes [24]. The LDA was computed with the plane-wave basis fixed at 480 eV with convergence measures which allowed energy changes per atom to be $<2 \times 10^{-5}$ eV. All atoms including hydrogen molecules in their respective atomic position were completely relaxed using the Broyden–Fletcher–Goldfarb–Shanno (BFGS) minimization methods which set the convergence tolerance less than 0.05 eV/Å, atom displacement less than 2×10^{-3} Å and stress less than 0.02 GPa. Additionally, the generalized gradient approximation (GGA)

MXenes: Fundamentals and Applications   Materials Research Forum LLC
Materials Research Foundations **51** (2019) 61-73   doi: https://doi.org/10.21741/9781644900253-3

method was selected to replace the LDA method in DFT coupling as it was more reliable in predicting hydrogen bonding in comparison with LDA methods [31]. Yadav et al. [25] employed GGA method for exchange-correlation functional determination for Cr_2C layers with similar plane-wave basis settings. Furthermore, the semi-empiral DFT-D2 method (Grimme method) is proposed as a coupled calculation method to predict the van der Waals interaction, while another coupled calculation method of DFT+U method was applied to obtain MXenes magnetic order [30].

To conduct a theoretical study on the hydrogen storage capacity of MXene material, the MAX phase of MXene had to be considered and constructed with their respective unit cell and optimized lattice parameters. The MXene constructed was then geometrically optimized for accurate computational value. The hydrogen storage capacity of the geometrically optimized MXene was determined with available sites for hydrogen atoms and molecules adsorption on both sides of the MXenes layer. For example in the study of Yadav et al. [25], the Cr_2C layer acted as the MXene layer had three sites considered for hydrogen storage capacity. The chromium atoms of the upper layer, carbon atoms of the middle layer and chromium atoms of the bottom layer as site-1, -2 and -3, respectively. The calculation of absorbed H-atom and H_2- molecule binding energy (E_b) on MXene layer was performed based on the Eq.1 where E_{host} is host structure total energy, $E_{H2/H}$ is free H_2/H total energy, $E_{host+nH2/H}$ is host structure adsorbed with new H_2/H total energy and n is new adsorbed H_2/H number.

$$E_b = (E_{host} + n\,E_{H2/H} - E_{host+nH2/H})/n \qquad (6)$$

The E_b result was used to determine the principal interaction of dissociated H-atom with MXene surface. Interactions such as strong chemical forces, weak physical forces, and Kubas-type interaction were present between the H-atom and MXene. The only desired interaction for hydrogen storage ability is Kubas-type interaction which allowed H-atoms bounded by MXene to be absorbed and released reversibly under ambient conditions. The Kubas-type interaction involved H_2 σ orbitals donating an electron to empty d orbitals of MXene while simultaneously, allowing the filled d orbitals of MXene to donate electron back to H_2 σ antibonding orbitals [24]. The calculated hydrogen storage ability was further compared with the gravimetric storage capacity target set at 5.5 wt% by the United States Department of Energy.

MXenes: Fundamentals and Applications   Materials Research Forum LLC
Materials Research Foundations **51** (2019) 61-73   doi: https://doi.org/10.21741/9781644900253-3

7.   Experimental study of Mxenes

A comparative study on hydrogen adsorption and desorption ability among several bare MXenes was conducted, and the result was shown in Fig. 2 [32]. An ideal MXenes should have a Gibbs free energy of hydrogen adsorption of atomic hydrogen (ΔG_{H*}^{a}) of approximately zero [33]. The positive ΔG_{H*}^{a} value denotes oxygen terminated MXenes which promoted hydrogen desorption while negative ΔG_{H*}^{a} value promoted hydrogen adsorption. For an ideal hydrogen storage material, the material should, therefore, have zero ΔG_{H*}^{a} value [33]. In their study, Ti_2CO_2 was located at the zero ΔG_{H*}^{a} value in the volcano curve and thus, this MXenes was expected to have the hydrogen storage ability among tested MXenes. From Fig. 2, the Ti_2C, Ti_3C_2, Nb_2C, V_2C, Nb_4C_3, V_2CO_2, and $Ti_3C_2O_3$ materials have low ΔG_{H*}^{a} value which favored hydrogen adsorption and would not able to act as a hydrogen storage material. Alternatively, Nb_2CO_2 and $Nb_4C_3O_2$ had a higher ΔG_{H*}^{a} value which preferred hydrogen desorption activities and couldn't act as hydrogen storage material too.

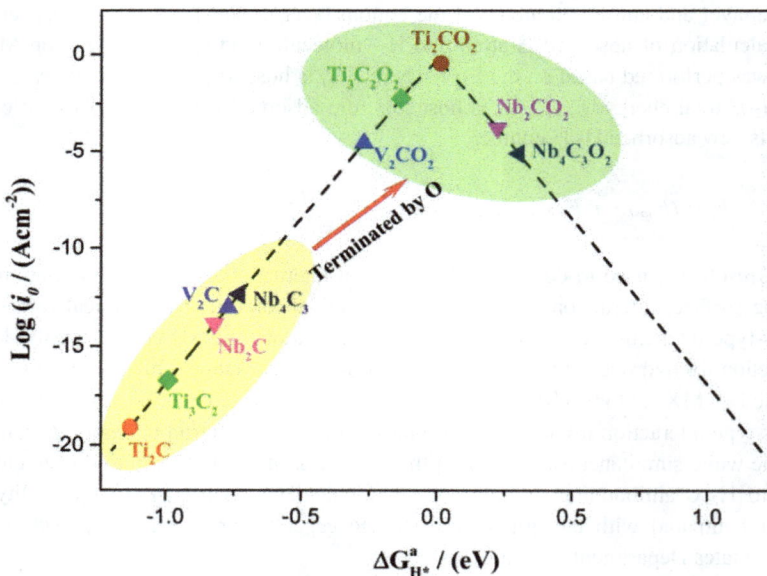

Figure 2: Volcano curve of hydrogen adsorption average Gibbs free energy in exchange current (i_o) function [32].

Bare MXenes were reported to have a lower reversible capacity at a particularly large current density which made it impractical for large scale hydrogen storage application [34]. Therefore, MXenes were introduced with different materials such as carbon nanotubes, sodium aluminium hydride (NaAlH$_4$) and external metal ions to increase their reversible capacity [34-36]. Experimental study on composite with NaAlH$_4$ and Ti$_3$C$_2$ MXene had enhanced hydrogenation/dehyrdrogenation kinetics at lower operation temperature [35]. The addition of NaAlH$_4$ onto the MXene reduced the activation energy for hydrogen desorption process which allowed reduction of operating temperatures. Another study on the improvement of Ti$_3$C$_2$ MXene was conducted with the introduction of MoS$_2$ and C nanohybrids [36]. The MoS$_2$/Ti$_3$C$_2$@C nanohybrids product had a superior high cycle life of 3000 cycles with a slow capacity loss of 0.0016% per cycl. The MOS$_2$ compound allowed fast charge transfer and high structural stability while carbon nanomaterial provided good conductivity, stability and compatibility for the MoS$_2$/Ti$_3$C$_2$-MXene product. The combination of all these compounds allowed enhancement of storage ability in MXene product.

Conclusion

As a novel 2–D layered material, MXene materials have received increasing enthusiasm in the past few years due to the distinctive properties that provided them great potential in a wide range of applications. The applications of MXenes garnered considerable interest, particularly in hydrogen storage. What is more, storing and releasing of H$_2$ at ambient conditions have played an important role to address the increasing demand for clean and recyclable energy. Properties such as ultra-large interlayer spacing, excellent biocompatibility, environmental benignity, hydrophilic surface property and good electric conductivity have prompted the use of MXenes towards H$_2$ storage. Whereas the preparation of MXenes the current chapter simply emphasized on the superficial parts of this area. The MXenes can be synthesized using an exfoliation method while their preparation was hard to obtain at normal temperature and pressure. With rapid advancement in this area, a timely account regarding the applications of MXenes in the hydrogen storage is extremely essential. We confidently believed that MXenes have extensive application prospects, especially for hydrogen storage and will be further discovered more comprehensively. We also anticipate that many scientists will spend great efforts in this field and bring substantial scientific developments in the near future.

Acknowledgments

Supports from Universiti Tunku Abdul Rahman (UTARRF/2018-C1/S02 and UTARRF/2018-C1/L02) and Ministry of Higher Education of Malaysia (FRGS/1/2016/TK02/UTAR/02/1) are greatly acknowledged.

MXenes: Fundamentals and Applications   Materials Research Forum LLC
Materials Research Foundations **51** (2019) 61-73   doi: https://doi.org/10.21741/9781644900253-3

References

[1]   L. Zhou, Progress and problems in hydrogen storage methods, Renew. Sust. Energ. Rev. 9 (2005) 395–408. https://doi.org/10.1016/j.apsusc.2018.04.052

[2]   D. Pukazhselvan, N. Nasani, T. Yang, D. Ramasamy, A. Shaula, D.P. Fagg, Chemically transformed additive phases in Mg_2TiO_4 and $MgTiO_3$ loaded hydrogen storage system MgH_2, Appl. Surf. Sci. 472 (2019) 99–104. https://doi.org/10.1016/j.apsusc.2018.04.052

[3]   E. Ianni, M.V. Sofianos, M.R. Rowles, D.A. Sheppard, T.D. Humphries, C.E. Buckley, Synthesis of $NaAlH_4$/Al composites and their applications in hydrogen storage, Int. J. Hydrog. Energy 43 (2018) 17309–17317. https://doi.org/10.1016/j.ijhydene.2018.07.072

[4]   K.K. Gangu, S. Maddila, S.B. Mukkamala, S.B. Jonnalagadda, Characteristics of MOF, MWCNT and graphene containing materials for hydrogen storage: A review, J. Energy Chem. 30 (2019) 132–144. https://doi.org/10.1016/j.jechem.2018.04.012

[5]   Y.T. Liu, X.D. Zhu, L. Pan, Hybrid architectures based on 2D MXenes and low-dimensional inorganic nanostructures: methods, synergies, and energy-related applications, Small 14 (2018) 1803632. https://doi.org/10.1002/smll.201803632

[6]   J. Pang, R.G. Mendes, A. Bachmatiuk, L. Zhao, H.Q. Ta, T. Gemming, H. Liu, Z. Liu, M.H. Rummeli, Applications of 2D MXenes in energy conversion and storage systems, Chem. Soc. Rev. 48 (2019) 72–133. https://doi.org/10.1039/c8cs00324f

[7]   N. Zhang, Y.H. Zhang, Y.J. Xu, Recent progress on graphene-based photocatalysts: current status and future perspectives, Nanoscale 4 (2012) 5792–5813. https://doi.org/10.1039/c2nr31480k

[8]   M. Naguib, M. Kurtoglu, V. Presser, J. Lu, J. Niu, M. Heon, L. Hultman, Y. Gogotsi, M.W. Barsoum, Two-dimensional nanocrystals produced by exfoliation of Ti_3AlC_2, Adv. Mater. 23 (2011) 4248–4253. https://doi.org/10.1002/adma.201102306

[9]   F. Wang, C. Yang, M. Duan, Y. Tang, J. Zhu, TiO_2 nanoparticle modified organ-like Ti_3C_2 MXene nanocomposite encapsulating hemoglobin for a mediator-free biosensor with excellent performances, Biosens. Bioelectron. 74 (2015) 1022–1028. https://doi.org/10.1016/j.bios.2015.08.004

[10]   H. Liu, C. Duan, C. Yang, W. Shen, F. Wang, Z. Zhu, A novel nitrite biosensor based on the direct electrochemistry of hemoglobin immobilized on MXene-Ti_3C_2, Sens. Actuators, B 218 (2015) 60–66. https://doi.org/10.1016/j.snb.2015.04.090

[11]   X.Q. Li, C.Y. Wang, Y. Cao, G.X. Wang, Functional MXenes materials: progress of their applications, Chem. Asian J. 13 (2018) 2742–2757.

[12]   J. Li, X.T. Yuan, C. Lin, Y.Q. Yang, L. Xu, X. Du, J.L. Xie, J.H. Lin, J.L. Sun, Achieving high pseudocapacitance of 2D titanium carbide (MXene) by cation intercalation and surface modification, Adv. Energy Mater. 7 (2017) 1602725. https://doi.org/10.1002/aenm.201602725

[13]   P. Persson, Exploring and tailoring structural properties of the two-dimensional family of Mxenes, European Microscopy Congress 2016: Proceedings, 2016, pp.437-438. https://doi.org/10.1002/9783527808465.emc2016.6942

[14]   J. Lei, X. Zhang, Z. Zhou, Recent advances in MXene: Preparation, properties, and applications, Front. Phys. 10 (2015) 276–286.

[15]   A. Enyashin, A. Ivanovskii, Structural and electronic properties and stability of MXenes Ti_2C and Ti_3C_2 functionalized by methoxy groups, J. Phys. Chem. C 117 (2013) 13637–13643. https://doi.org/10.1021/jp401820b

[16]   G. Chen, Y. Zhang, H. Cheng, Y. Zhu, L. Li, H. Lin, Effects of two-dimension MXene Ti_3C_2 on hydrogen storage performances of MgH_2-$LiAlH_4$ composite, Chem. Phys.522 (2019) 178-187. https://doi.org/10.1016/j.chemphys.2019.03.001

[17]   M. Wu, B.X. Wang, Q.K. Hu, L.B. Wang, A.G. Zhou, The synthesis process and thermal stability of V_2C MXene, Materials 11 (2018) 2112. https://doi.org/10.3390/ma11112112

[18]   M. Alhabeb, K. Maleski, B. Anasori, P. Lelyukh, L. Clark, S. Sin, Y. Gogotsi, Guidelines for synthesis and processing of two-dimensional titanium carbide ($Ti_3C_2T_x$ MXene), Chem. Mater. 29 (2017) 7633–7644. https://doi.org/10.1021/acs.chemmater.7b02847

[19]   O. Mashtalir, M. Naguib, V.N. Mochalin, Y.D. Agnese, M. Heon, M.W. Barsoum, Y. Gogotsi, Intercalation and delamination of layered carbides and carbonitrides, Nat. Commun. 4 (2013) 1716. https://doi.org/10.1038/ncomms2664

[20]   F. Chang, C.S. Li, J. Yang, H. Tang, M.Q. Xue, Synthesis of a new graphene-like transition metal carbide by de-intercalating Ti_3AlC_2, Mater. Lett. 109 (2013) 295–298. https://doi.org/10.1016/j.matlet.2013.07.102

[21]   M. Naguib, J. Halim, J. Lu, K.M. Cook, L. Hultman, Y. Gogotsi, M.W. Barsoum, New two-dimensional niobium and vanadium carbides as promising materials for Li-ion batteries, J. Am. Chem. Soc. 135 (2013) 15966–15969. https://doi.org/10.1021/ja405735d

[22]   M. Naguib, O. Mashtalir, J. Carle, V. Presser, J. Lu, L. Hultman, Y. Gogotsi, M.W. Barsoum, Two-dimensional transition metal carbides, ACS Nano 6 (2012) 1322–1331. https://doi.org/10.1021/nn204153h

[23]   J. Halim, M.R. Lukatskaya, K.M. Cook, J. Lu, C.R. Smith, L.A. Naslund, S.J. May, L. Hultman, Y. Gogotsi, P. Eklund, M.W. Barsoum, Transparent conductive two-dimensional titanium carbide epitaxial thin films, Chem. Mater. 26 (2014) 2374–2381. https://doi.org/10.1021/cm500641a

[24]   Q. Hu, D. Sun, Q. Wu, H. Wang, L. Wang, B. Liu, A. Zhou, J. He, MXene: A new family of promising hydrogen storage medium, J. Phys. Chem. A 117 (2013) 14253–14260. https://doi.org/10.1021/jp409585v

[25]   A. Yadav, A. Dashora, N. Patel, A. Miotello, M. Press, D. Kothari, Study of 2D MXene Cr_2C material for hydrogen storage using density functional theory, Appl. Surf. Sci. 389 (2016) 88–95. https://doi.org/10.1016/j.apsusc.2016.07.083

[26]   T.K.A. Hoang, D.M. Antonelli, Exploiting the Kubas interaction in the design of hydrogen storage materials, Adv. Mater. 21 (2009) 1787–1800. https://doi.org/10.1002/adma.200802832

[27]   S.I. Orimo, Y. Nakamori, J.R. Eliseo, A. Züttel, C.M. Jensen, Complex hydrides for hydrogen storage, Chem. Rev. 107 (2007) 4111–4132. https://doi.org/10.1021/cr0501846

[28]   J.L. Rowsell, O.M. Yaghi, Strategies for hydrogen storage in metal-organic frameworks, Angew. Chem. Int. Ed. 44 (2005) 4670–4679. https://doi.org/10.1002/anie.200462786

[29]   Q.K. Hu, H.Y. Wang, Q.H. Wu, X.T. Ye, A.G. Zhou, D.D. Sun, L.B. Wang, B.Z. Liu, J.L. He, Two-dimensional Sc_2C: A reversible and high-capacity hydrogen storage material predicted by first-principles calculations,Int. J. Hydrog. Energy 39 (2014) 10606–10612. https://doi.org/10.1016/j.ijhydene.2014.05.037

[30]   M. Khazaei, A. Ranjbar, M. Arai, T. Sasaki, S. Yunoki, Electronic properties and applications of MXenes: a theoretical review, J. Mater. Chem. C 5 (2017) 2488–2503. https://doi.org/10.1039/c7tc00140a

[31]   T.J. Giese, D.M. York, Density-functional expansion methods: Evaluation of LDA, GGA, and meta-GGA functionals and different integral approximations, J. Chem. Phys. 133 (2010) 244107. https://doi.org/10.1063/1.3515479

[32]   G.P. Gao, A.P. O'Mullane, A.J. Du, 2D MXenes: A New Family of Promising Catalysts for the Hydrogen Evolution Reaction, ACS Catal. 7 (2017) 494–500. https://doi.org/10.1021/acscatal.6b02754

[33]   N.K. Chaudhari, H. Jin, B.Y. Kim, D.S. Baek, S.H. Joo, K.Y. Lee, MXene: an emerging two-dimensional material for future energy conversion and storage applications, J. Mater. Chem. A 5 (2017) 24564-24579. https://doi.org/10.1039/c7ta09094c

[34]   S.J. Sun, C. Liao, A.M. Hafez, H.L. Zhu, S.P. Wu, Two-dimensional MXenes for energy storage, Chem. Eng. J. 338 (2018) 27–45. https://doi.org/10.1016/j.cej.2017.12.155

[35]   R.Y. Wu, H.F. Du, Z.Y. Wang, M.X. Gao, H.G. Pan, Y.F. Liu, Remarkably improved hydrogen storage properties of NaAlH$_4$ doped with 2D titanium carbide, J. Power Sources 327 (2016) 519–525. https://doi.org/10.1016/j.jpowsour.2016.07.095

[36]   X.H. Wu, Z.Y. Wang, M.Z. Yu, L.Y. Xiu, J.S. Qiu, Stabilizing the MXenes by carbon nanoplating for developing hierarchical nanohybrids with efficient lithium storage and hydrogen evolution capability, Adv. Mater. 29 (2017) 1607017. https://doi.org/10.1002/adma.201607017

Materials Research Forum LLC
doi: https://doi.org/10.21741/9781644900253-4

Chapter 4

MXenes for Electrocatalysis

Wenyu Yuan[1,2]*, Laifei Cheng[1]

[1]Science and Technology on Thermostructural Composite Materials Laboratory, Northwestern Polytechnical University, Xi'an 710072, China

[2]Global Research Center for Environment and Energy based on Nanomaterials Science (GREEN), National Institute for Materials Science (NIMS), Tsukuba 305-0044, Japan

YUAN.Wenyu@nims.go.jp, ywyonly1992@hotmail.com

Abstract

Electrocatalysis is considered as one of the most promising approaches to overcome the global energy crisis and environmental challenges. The high-performance electrocatalytic system heavily relies on the design of catalysts. MXenes, a large family of novel 2D transition metal carbides (nitrides), which is obtained from selectively etching of MAX ceramics, are candidates for electrocatalysis. In this chapter, we will mainly introduce the advances of MXene-based electrocatalysts, and discuss the applications of MXenes in hydrogen evolution reaction (HER), oxygen evolution reaction (OER), and N_2 reduction reaction (NRR). Finally, the future outlook of MXenes in electrocatalysis will analyzed in-depth.

Keywords

MXene, Electrocatalysis, HER, OER, NRR

Contents

1.   Introduction

The global energy & environmental crisis claims for novel green energy generation technologies [1]. Electrocatalysis, a technology that can convert molecules in the atmosphere (e.g., H_2O, CO_2, and N_2) into higher-value products (e.g., H_2, CH_3OH, CO, and NH_3), is considered as a promising approach to overcome these challenges, and recently attracted growing attentions [2]. For example, electrocatalytic CO_2 reduction is a candidate technology to slow down global warming [3]. Electrocatalytic NRR can significantly reduce current energy consumption because ~1-2% of the global energy is used for the fixation of N_2 via traditional high temperature reduction. Based on electrocatalysis, a sustainable energy future can be built. In this chapter, we will mainly introduce three electrocatalytic technologies: HER, OER, and N_2 fixation.

The electrocatalytic performance heavily relies on electrocatalysts, which play key roles in efficiency, activity, and selectivity. Most of the noble metals and their oxides possess high activity and efficiency towards electrocatalysis [2]. However, the high-cost and poor abundance limit their wide applications. Therefore, searching for highly active non-noble metal based electrocatalysts with low-cost become a hot topic in recent years. To date, various electrocatalysts have been reported as highly active catalysts, including metal carbides (e.g., Mo_2C, WC), metal nitrides (e.g., TiN, Co_4N, MoN), metal oxides (e.g., Co_3O_4, Fe_2O_3), and metal hydroxides (e.g., $Ni(OH)_2$, layered double hydroxide) [4, 5]. Although the non-noble metal-based catalysts' performance has been significantly enhanced via various modification approaches, the quest to seek novel electrocatalysts is still on its way.

MXene, as a new kind of large 2D-transition metal carbides family, which was discovered in 2011, has gained widespread attention as potential highly active electrocatalysts [6-9]. The unique 2D structures of MXenes provide fast ion/charge transfer path and are beneficial to surface chemical reactions. To date, various electrocatalytic properties of MXenes, including HER, OER, CO_2 reduction, NRR, etc., have been reported, suggesting great potentials in electrocatalytic fields. In this chapter, we will focus on the HER, OER and NRR performance of MXenes. The mechanism of

MXenes: Fundamentals and Applications   Materials Research Forum LLC
Materials Research Foundations 51 (2019) 74-104   doi: https://doi.org/10.21741/9781644900253-4

these electrocatalytic technologies will be detailed, and the MXene-based electrocatalysts will be discussed.

2.   MXenes forHER

2.1   The mechanism of HER

H_2 is considered as one of the promising candidates for clean-energy carriers for replacing nonrenewable, environmental-unfriendly fossil fuels owing to its high energy density (140 kJ/g) which is three times that of gasoline, renewability, and environmental benignity [10]. However, most of the hydrogen at present is mainly produced from fossil fuels, leading to high cost, high greenhouse gas emission, and low efficiency, thus accelerating the development of more energy efficient and cost-effective methods for H_2 generation [11]. Among these H_2 generation methods, water splitting is one of the most attractive, efficient, and simple methods to produce H_2.

HER, an important technology via electrochemical reduction of water to generate H_2 at the cathodic side, attracts extensive interest due to the high energy conversion efficiency. The potential to generation H_2 via HER can be expressed as [12]:

$$E_i = E_{HER} + iR + \eta \tag{1}$$

where E_{HER} is the Nernstian potential for the HER (0 V $v.s.$ RHE), iR represents for the ohmic potential drop and η represents for the reaction overpotential. In fact, overpotential is a very important factor to evaluate the electrocatalytic performance. Usually, the high electrocatalytic activity requires low overpotential. The overpotential can be calculated by the following equation [12, 13]:

$$\eta = a + b \log j = \frac{-2.3RT}{\alpha nF} \log j_0 + \frac{2.3RT}{\alpha nF} \log j \tag{2}$$

Where α represents for the charge transfer coefficient, n represents for the number of electrons transferred, R represents for the ideal gas constant, and T represents for the temperature, F represents for the Faraday constant, j_0 represents for the exchange current density, and j represents for the current density. The overpotential is linearly correlated with $\log j$.

The slop $2.3RT/\alpha nF$ is defined as the Tafel slope, which is usually used to evaluate the rate-determining step and study the whole electrocatalytic reaction mechanism. In

MXenes: Fundamentals and Applications   Materials Research Forum LLC
Materials Research Foundations **51** (2019) 74-104   doi: https://doi.org/10.21741/9781644900253-4

general, the reaction pathways consist of two steps: Volmer reaction, Heyrovsky or Tafel reaction. The first step, Volmer reaction, is the combination of one proton from the media and one electron from the catalyst, and resulting in the adsorption of hydrogen on the surface of the catalyst. The reactions in alkaline media demand additional energy to water dissociation, and slowing down the reaction rate and affecting the reaction equation. The reaction equation can be expressed as [14]:

$$H^+ + e^- + * \rightarrow H* \text{ (Volmer) acidic electrolytes} \tag{3}$$

$$H_2O + e^- \rightarrow H* + OH^- \text{ (Volmer) alkaline electrolytes} \tag{4}$$

$$b = \frac{2.3RT}{\alpha F} \approx 120 \, \text{mVdec}^{-1} \tag{5}$$

After the first step, two possible reactions to generate H_2 are subsequently followed. One possible way is that the adsorbed hydrogen atom by Volmer reaction combines with one proton and one electron, and then generates molecular H_2. This step is named as Heyrovsky reaction:

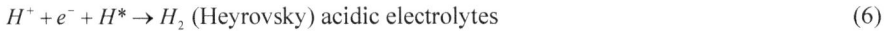

$$H^+ + e^- + H* \rightarrow H_2 \text{ (Heyrovsky) acidic electrolytes} \tag{6}$$

$$H_2O + e^- + H* \rightarrow H_2 + OH^- \text{ (Heyrovsky) alkaline electrolytes} \tag{7}$$

$$b = \frac{2.3RT}{(1+\alpha)F} \approx 40 \, \text{mVdec}^{-1} \tag{8}$$

The other possible way is that two absorbed hydrogen combines to molecular H_2, known as the Tafel reaction:

$$2H* \rightarrow H_2 \text{ (Tafel)} \tag{9}$$

$$b = \frac{2.3RT}{2F} \approx 30 \, \text{mVdec}^{-1} \tag{10}$$

The rate of hydrogen evolution is mainly influenced by the process of absorbing hydrogen to the surface of catalysts, known as the free energy of hydrogen absorption (ΔG_H). Parson firstly pointed out that when $\Delta G_{H*} = 0$, the maximum activity can be achieved in 1958. If the absorption capability is too weak, the adsorption step would be the rate-determining step. If the absorption capability is too strong, the reaction−desorption step would be the rate-determining step. Only when $\Delta G_{H*} = 0$, the absorption capability is neither too strong nor too weak. Therefore, a volcano plot to describe the catalytic activity and ΔG_{H*} can be obtained to study the actvitiies of catalysts. Figure 1ashows a typical volcano plot to describe the relationship between $j_0$and ΔG_{H*}[15]. To enhance the activity of HER, a catalyst with minimum ΔG_H and modified surface must be explored.

Beyond the theoretical calculations, from the discussion above, the overpotential, Tafel slope, and j_0 are key indexes for the evaluation of the activity for HER reaction. The most widely accepted technologies to measure the HER reaction activity is the linear sweep voltammetry (LSV), and the common results are illustrated in Figure 1b-1c. Based on the LSV data, the plot of E-log j (Tafel plot) can be obtained. The onset overpotential is determined by the potential when the E starts to linearly increase with the change of log j under low current densities. Under high current densities, the Tafel plot shows a linear relationship and the slop is the Tafel slop. The point of intersection of the slop and x-axis is the log j_0 [16]. Therefore, the exchange current density (j_0) can be obtained.

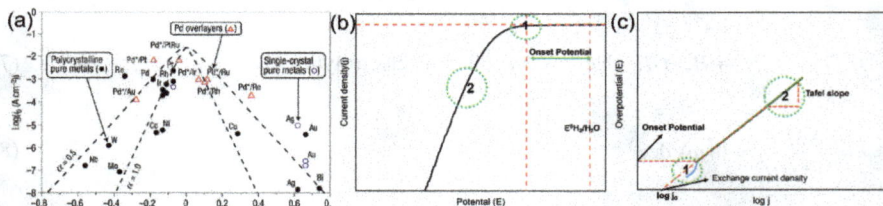

Figure 1. (a) Typical volcano plot to describe the HER activities of various metals and metal-based hybrids. The ΔG_{H*} values here are calculated under room temperature (25°C) with a hydrogen coverage of either 1/4 or 1/3. The αhere represents transfer coefficients[15].(b) Schematic HER polarization curves. (c) Schematic Tafel plots.

To date, precious metals (e.g., Pt) based catalysts have the highest activities among these catalysts. However, the limited capacity on earth, high cost, and non-renewability of precious metals make them far from industrialized applications, thus accelerating the development of non-precious metal-based catalysts, such as various metal carbides, nitrides, and their hybrids. These catalysts, on the one hand, can significantly reduce the costs of catalysts. On the other hand, they suffer from incomparable activity to precious metal catalysts. Seeking for novel active catalysts for HER is extremely urgent.

2.2 MXene-based catalysts for HER

Owing to their outstanding electrical conductivities, large specific surface areas (SSA) and unique 2D structures, MXenes are considering as potential catalysts for HER. Gogotsi's group [17] for the first time reported the HER performance of MXenes in the year of 2016. Both theoretical and experimental results suggested that the HER activity of 2D-Mo_2C was greatly higher than that of Ti_2C. The overpotential of Mo_2C was highly reduced to 0.283 V, while that of Ti_2C reached an overpotential of >0.6 V. This research facilitates the HER performance study of MXenes.

DFT calculations proved that O-terminated MXenes are potential catalysts for HER which are candidates to replace Pt. Aijun Du *et al.* [18] pointed out that Ti_2C, V_2C, Nb_2C, and Ti_3C_2 are electrically conductive under standard conditions, indicating the high charge transfer kinetics of these MXene materials for HER, as shown in Figure 2a [18]. Remarkably, the ΔG_{H*} of MXenes terminated via O atoms is close to 0 eV, which is much lower than other 2D materials (e.g., MoS_2, WS_2), indicating a Pt-like high activity (Figure 2b)[19]. Jinlan Wang *et al.* [20] investigated ~10 kinds of metal carbides, and a volcano plot between Ne and ΔG_{H*} was obtained, as shown in Figure 2c [20]. Ti_2VCO_2 is the highest active catalyst among all investigated MXene carbides with a low ΔG_{H*} of 0.01~0.06 eV. Jinlan Wang *et al.* [21] recently found that Ti_2NO_2 and Nb_2NO_2 are also novel promising HER electrocatalysts with an extremely low ΔG_{H*} (close to 0 eV) [21]. Besides, low coverage of fluorine on the basal plane of MXenes usually benefits for the improvement of HER activity [22]. Furthermore, recently, 2D ordered double $M'_2M''C_2$ terminated with oxygen atoms (e.g., $Cr_2VC_2O_2$, $Cr_2TiC_2O_2$, $Mo_2VC_2O_2$, $Mo_2TiC_2O_2$) are highly efficient catalysts for HER with low ΔG_{H*} of ~ 0 eV (Figure 2d) [23]. The oxygen atoms in the surface of $M'_2M''C_2O_2$ which offered suitable interaction strength between oxygen terminated MXenes and hydrogen were defined as the active sites for HER. Owing to the unique electronic structure, more electrons can be obtained by the terminated O^* from $M'_2M''C_2$. The suitable interaction strength between terminated oxygen and absorbed hydrogen was caused by the high occupation of the p-orbitals of the terminated oxygen atoms.

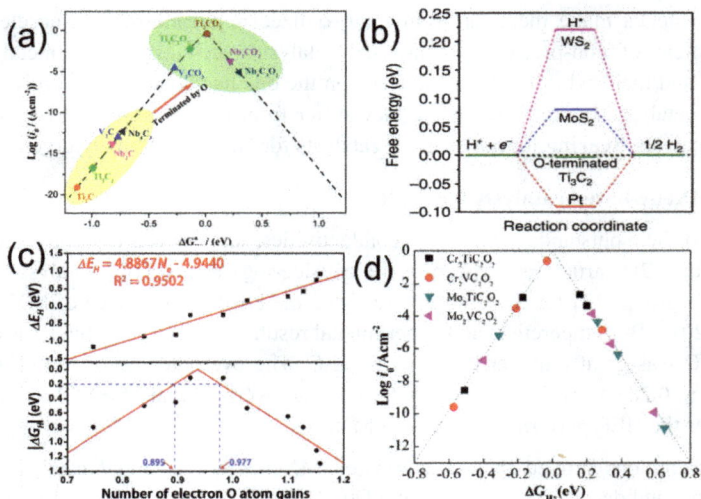

Figure 2. (a) Volcano curve of j_0 as a function of ΔG_{H*}[18].(b) ΔG_{H*} of O-terminated Ti_3C_2, and MoS_2, Pt and WS_2[19]. (c) The calculated ΔE_H (upper) and $|\Delta G_{H*}|$ (down) v.s. the number of electron O atom gains (Ne) [20]. (d) Volcano plot of j_0 as a function of ΔG_{H*}. for various double metal carbides (MXenes)[23].

Pure V_2CO_2 is not an ideal catalyst for HER. Jinlan Wang *et al.* found that transition metal anchored V_2CO_2 is a promising candidate for HER via DFT calculations [24]. Via adjusting the suitable coverage and type of the promoters, and choosing suitable active sites, the optimal ΔG_{H*} value can be reduced to ~0 eV. Interestingly, strain engineering can modulate the active sites. Experimental works also demonstrated vanadium-based MXenes are active catalysts. It was recently also reported as a potential HER catalysts [25]. The current density could reach 10 mA cm^{-2}for V_4C_3 under a low overpotential of ~0.2 V.

The seeking for novel MXene-based HER catalysts is also an important research target. M_3CNO_2, which is another large group in the MXene family, recently has been investigated as HER catalysts [26]. The DFT calculations suggest that Ti_3CNO_2 and Nb_3CNO_2 are potential HER catalysts among M_3CNO_2 MXenes. The activities can be further enhanced via using O/OH mixed terminated atoms, and transition metal modifications, which is attributed by that O atoms gaining more electrons after modifications. 2D metal borides recently were also reported to be efficient electrocatalysts for HER [27]. 2D Fe_2B_2 showed extremely small ΔG_{H*} of ~0 eV (−0.007, −0.001, 0.038, and 0.059 eV for the H* coverage of 1/8, 1/4, 3/8, and 1/2, respectively), in which all Fe atoms in-plane are active sites for HER.

To promote HER activity, various approaches have been carried out to expose more active sites, improve the electrical conductivity and modify the electronic structure. Yuan *et al.* [28] designed a $Ti_3C_2T_x$ nanofiber catalysts via the hydrolyzation of bulk Ti_3AlC_2 ceramics and the selectively etching of "Al" phase, in which the OH⁻ worked like a scissor to create a crack in bulk Ti_3AlC_2 ceramics, and the shear force led to the avulsion of bulk Ti_3AlC_2, and leading to the generation of Ti_3AlC_2 nanofibers (shown in Figure 3a) [28]. The MXene nanofibers with large SSA, fast electron transfer path, and large amounts of active sites, significantly enhanced the activity of MXenes. A current density of 10 mA cm^{-2} can be reached under a low potential of -169 mV for $Ti_3C_2T_x$ nanofibers. The obtained Ti_3C_2 nanofibers were more likely to be nanoribbons, and the results were further supported via DFT calculations [29]. The DFT calculations suggested that the hydrogen can be quickly absorbed on the edges of MXene nanoribbons which are considered as the active sites for HER. Among all investigated MXene nanoribbons, Ti_3C_2 nanoribbon provides an extremely low $\triangle G_{H*}$ of -0.07 eV at a hydrogen coverage of 0.5, is the most active catalyst. Ki-Seok An *et al.* [30] found that the chemical bonding of Ti–N$_x$ on the surface of Ti_2CT_x can highly promote HER activity (Figure 3c) [30]. When the current density reached 10 mA cm^{-2}, the overpotential is only 0.215 V, which is much lower than that of Ti_2CT_x (0.645 V). The $\triangle G_{H*}$ after the nitrogen-doping was also reduced (Figure 3d-e). A similar approach to modify the Ti_3C_2 was also reported by Yuan *et al.* [31] via one-step nitridation (Figure 3b) [31]. Zhou *et al.* [32] also demonstrated that nitrogen-doped MXenes (N-Mo_2C) could expose more active Mo atoms, leading to high HER activity. The reported N-Mo_2C delivered an extremely low onset-overpotential of 48.3 mV, a small Tafel slope of 44.5 mV dec^{-1} and potential of only -99 mV (at 10 mA cm^{-2}), which is comparable to Pt/C catalysts. Qingyu Yan *et al.* [33] reported that nitrogen and sulfur co-doped Mo_2C could also highly improve the HER activity, in which the doping can highly enhance the wettability with an aqueous solution and induce the synergistic effects.

MXene-based hybrid structures are also efficient methods to promote their activity owing to the synergistic effects. Up till now, MXene-metal oxides, MXene-metal sulfides, MXene-metal organic frameworks (MOFs), and other MXene hybrids have been reported for HER. Jijun Zhao *et al.* [34] carried out DFT calculations on MXene/nitrogen-doped graphene and found that the graphiticsheet/V_2C or Mo_2C MXenes is highly active catalysts for HER with a low reaction free energyof~0 eV. The electronic coupling between the graphene and MXene were responsible for the outstanding catalytic activities

(Figure 4a). Loh *et al.* [35] reported a hybrid catalyst of 2D-Mo_2C and graphene, which is synthesized via chemical vapor deposition (CVD) process, and showed high HER activity with a low potential of -87 mV at the current density of 10 mAcm^{-2}.Yan *et al.* [36] demonstrated a MXene/$Ni_xF_{1-x}S_3$ hybrid catalysts, and the performance can be controlled by adjusting the Ni/Fe ratio. After optimizing, the$Ni_{0.9}Fe_{0.1}PS_3$@MXene shows low overpotential of 196 mV for HER in 1 M KOH solution. Recently, Chun Cheng *et al.* [37] designed a molybdenum nitrides (MXenes) embedded in boron, nitrogen codoped-carbonnanotubes (MoN/BN-CNT) structure for HER. The synthesized MoN/BN-CNT catalyst delivered an extremely small overpotential of 78 mV at 10 mA cm^{-2}, meanwhile the Tafel slope was only 46 mV dec^{-1}. Besides, the obtained catalysts can be operated in large pH-windows, even seawater.

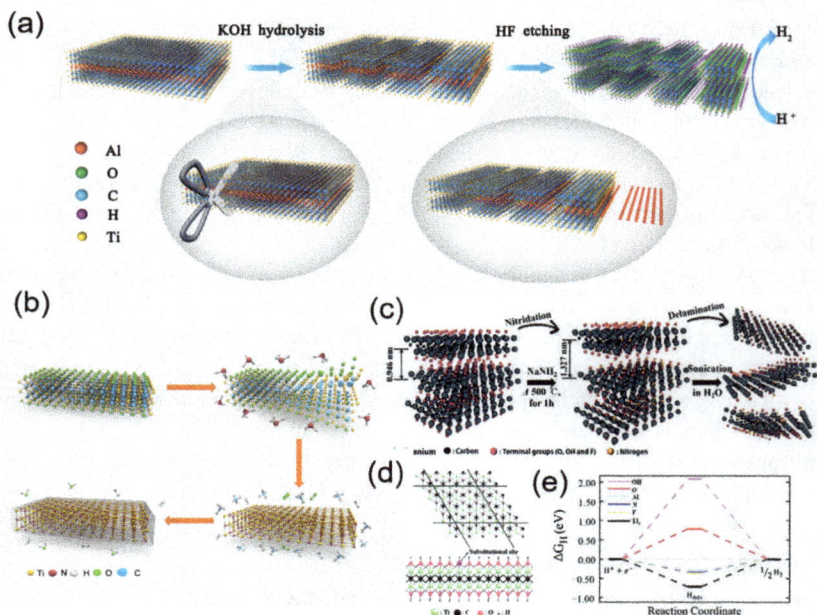

Figure 3. (a) The schematic diagram for the synthesis of Ti_3C_2 nanofibers [28]. (b) The one-step nitridation mechanism of Ti_3C_2 [31]. (c) The N-doping modification for $Ti_3C_2T_x$ (d) the supercell strucuture of nitrogen-doped Ti_3C_2 and (e) the corresponding ΔG_{H*} [30].

$MoS_2@Ti_3C_2$ is also a potential HER catalyst, which is proved by Daniel R. Strongin *et al.* [38]. The interlayer spacing of MXenes was expanded owing to the 3D-vertically aligned MoS_2 nanosheets. The $MoS_2@Ti_3C_2$ catalyst exhibited a low onset overpotential of ~95 mV to activate the HER and a low overpotential of ~110 mV (at 10 mA cm^{-2}) with a low Tafel slope of 40 mV dec^{-1}. Considering the poor oxygen-resistance of MXenes, Jieshan Qiu *et al.* [39] demonstrated a carbon nanoplating strategy to stabilize MXenes during the hydrothermal process. The obtained $MoS_2/Ti_3C_2@C$ possessed high stability and HER catalytic activity in acidic solution. Impressively, $MoS_2/Ti_3C_2@C$ shows smallest onset overpotential of ~20 mV among all reported MXene based HER catalysts. Apotential of -135 mV was required to reach a current benchmark density of 10 mA cm^{-2} with a small Tafel slope of 45 mV dec^{-1}. The Ti_3C_2/MoS_2' s HER activity can be furthered promoted via structural design. Fan *et al.* [40] designed an anoroll-like Ti_3C_2/MoS_2 hybrid by combining liquid N_2-freezing and a subsequent annealing process. The obtained Ti_3C_2/MoS_2 catalyst exhibited a small onset overpotential of 30 mV, and the j_0 of Ti_3C_2/MoS_2 was ~ 25 times that of MoS_2. Jieshan Qiu *et al.* [41] recently reported a 3D-MXene@CoP structure via 3D capillary-forced assembly process (Figure 4b-g)[41]. 3D-MXene can highly render high robustness and excellent processability. Therefore, the $CoP@3D-Ti_3C_2$ catalyst delivered an overpotential of ~168 mV (at 10 mA cm^{-2}) in 1M KOH, which is comparable to commercial Pt/C catalyst, and superior to MXene-freecatalysts in alkaline electrolyte.

The theoretical study demonstrated that the HER activities of MXenes could be modified by transition metal atoms (Figure 4h)[42]. The presence of transition metal not only optimizes the ΔG_{H*} but also reduces the activation barrier. The HER mechanism was also switched from Volmer-Heyrovsky mechanism for MXenes to Volmer-Tafel for MXene modified by transition metal atoms. Liwei Lin *et al.* [43] demonstrated that Mo_2C-Co hybrids are active catalysts for HER. A remarkably small overpotential of 48 mV was delivered under a current density of 10 mA cm^{-2}, meanwhile the Tafel slope was only 39 mV dec^{-1}, which is comparable to Pt/C catalysts. Very recently, single-atom/MXene hybrids were proved to be efficient catalysts for HER. Yadong Li, Guoxiu Wang, and Yury Gogotsi *et al.* [44] reported a composite catalyst of $Mo_2TiC_2T_x$ and single Pt atoms for HER via a simple electrochemical process, as shown in Figure 5 [44]. The $Mo_2TiC_2T_x$ with large amounts of exposed (0001) facets and the abundant Mo vacancies exposed in the outer layers were used for the fixation of single atomic- Pt, and enhance the MXene's catalytic activity. As a result, the $Mo_2TiC_2T_x$ with atomic Pt exhibited ultrahigh catalytic ability. The overpotentials to achieve 10 mA cm^{-2} was reduced to 30 mV, meanwhile the Tafel slope was suppressed to only 30 mV dec^{-1} in acidic solution. The mass activity was 39.5 times greater than that of commercial Pt/C catalyst. The calculated ΔG_{H*} of obtained

$Mo_2TiC_2O_2$–Pt_{SA} is also lower than Pt and $Mo_2TiC_2O_2$, suggesting the high HER activity of MXene-single atomic metal hybrids. Similarly, Junliang Sun *et al.* [45] synthesized an MXene/Pt catalyst via a wet-impregnation and photo-induced reduction method, and the obtained catalyst achieved a low overpotential of 55 mV at a current density of 10 mA cm^{-2} [45]. Pt_xNi ultrathin nanowires/Ti_3C_2 were efficient HER catalysts [46]. Specifically, $Pt_{3.21}Ni/Ti_3C_2$ achieved ultrahigh performance with low potential (-18.55 mV) and small Tafel slope (13.37 mV dec^{-1}) in acidic media.

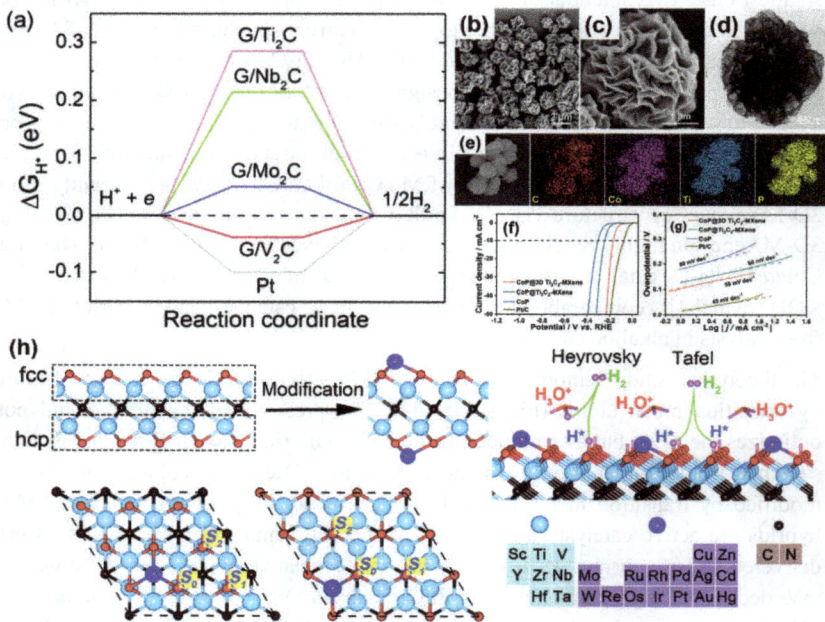

Figure 4. (a) the $\triangle G_{H*}$ of the HER on the most active sites of graphene/MXene heterostructures [34]. (b-g) The morphologies and HER performance of 3D MXene/CoP hybrids [41]. (h) Schematic diagrams of the modification of transition metal atoms on O-terminated MXenes [42].

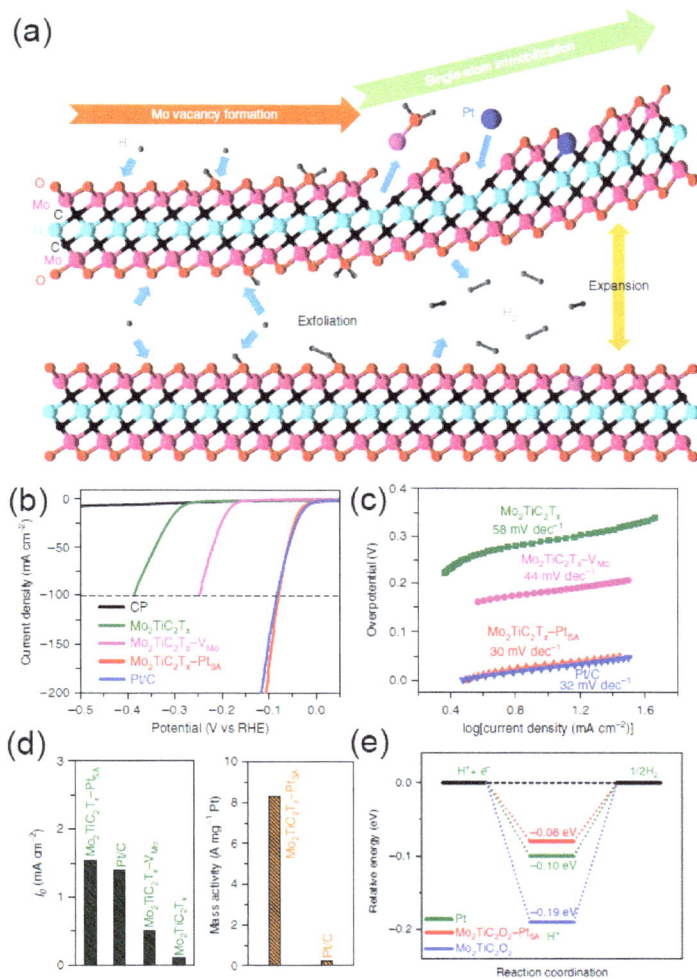

Figure 5. (a) Illustration of the synthesis mechanism for $Mo_2TiC_2O_2$–Pt_{SA} during the HER process. (b) HER polarization curves of carbon paper (CP), $Mo_2TiC_2T_x$, $Mo_2TiC_2T_x$–V_{Mo}, $Mo_2TiC_2T_x$–Pt_{SA} and Pt/C (40%). The graphite rod was used as the counter electrode. (c) Corresponding Tafel slope derived from Figure 5a. (d) j_0 of the catalysts, and the mass activity of commercial Pt/C and $Mo_2TiC_2T_x$–Pt_{SA}. (e) Calculated ΔG_{H*} of HER at the equilibrium potential for $Mo_2TiC_2O_2$, $Mo_2TiC_2O_2$–Pt_{SA} and Pt/C [44].

To clearly show the research progress of MXene-based catalysts for HER, we summarized the latest MXene-based hybrids and their HER performance in Table 1. On the one hand, MXenes can serve as catalysts directly. On the other hand, MXenes are also an excellent substrate to enhance other catalysts' performance. The noble metal-MXene hybrid catalysts delivered highest HER activities. However, the use of these noble metals is not practical owing to their high cost and scarcity. Therefore, designing highly efficient non-noble metal-MXene catalysts, seeking novel MXene-based catalysts, and the modification of MXenes are still of critical significance for the development of HER.

Table 1. MXene-based catalysts for HER

Materials	Solution	Mass loading (mg cm^{-2})	Overpotential (mV at 10 mA cm^{-2})	Tafel slope (mV dec^{-1})	References
Ti_2C	0.5 M H_2SO_4	0.1	609	124	[17]
Mo_2C	0.5 M H_2SO_4	0.1	283	82	[17]
Mo_2C/Graphene	0.5 M H_2SO_4	Not given	87	73	[35]
Ti_3C_2 nanofiber	0.5 M H_2SO_4	~0.3	169	97	[28]
Ti_3C_2/MoS_2	0.5 M H_2SO_4	~0.7	110	40	[40]
N-Mo_2C	0.5 M H_2SO_4	~0.36	99	44.5	[32]
N,S-Mo_2C	0.5 M H_2SO_4	~0.36	86	47	[33]
Ti_3C_2/$Ni_xF_{1-x}S_3$	1 M KOH	0.25	196	114	[36]
MoN/BN-CNT	0.5 M H_2SO_4	~0.25	78	46	[37]
MoS_2/Ti_3C_2@C	0.5 M H_2SO_4	0.4	135	45	[41]
Mo_2C/Co	0.5 M H_2SO_4	1	48	39	[43]
Mo_2TiC_2 /Pt	0.5 M H_2SO_4	0.5	30	30	[44]
$Pt_{3.21}Ni$/Ti_3C_2	0.5 M H_2SO_4	~70	18.6	13.4	[46]
Ti_3C_2/CoP	1 M KOH	0.2	168	58	[38]

3.   MXene for OER

3.1   The mechanism of OER

OER, an electrocatalytic oxidation process of water to molecular oxygen, is a key reaction for water splitting. OER suffer from a four electron-proton coupled reaction, and high activation energy for the formation of O=O bond, while only two electron-transfer is required for HER. Hence, OER requires higher energy to overcome the kinetic barrier [47]. The OER's reaction equation can be drawn as follows [48]:

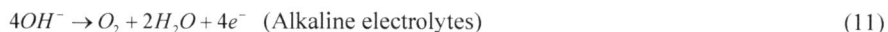

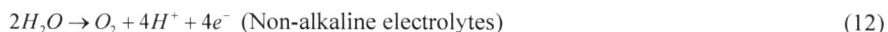

$$4OH^- \rightarrow O_2 + 2H_2O + 4e^- \quad \text{(Alkaline electrolytes)} \tag{11}$$

$$2H_2O \rightarrow O_2 + 4H^+ + 4e^- \quad \text{(Non-alkaline electrolytes)} \tag{12}$$

To date, many possible mechanisms for OER have been proposed by different research groups, and some of them are different. Here, the possible mechanism which is widely accepted is given here and shown in Figure 6 [49]. The reactions are given as follows:

In non-alkaline electrolytes [50]:

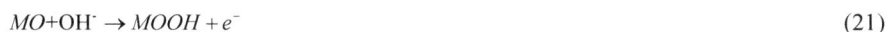

$$M + H_2O \rightarrow MOH + H^+ + e^- \tag{13}$$

$$MOH + OH^- \rightarrow MO + H_2O + e^- \tag{14}$$

$$2MO \rightarrow 2M + O_2(g) \tag{15}$$

$$MO + H_2O(l) \rightarrow MOOH + H^+ + e^- \tag{16}$$

$$MOOH \rightarrow M + O_2(g) + H^+ + e^- \tag{17}$$

In alkaline electrolytes:

$$M + OH^- \rightarrow MOH + e^- \tag{18}$$

$$MOH + OH^- + e^- \rightarrow MO + H_2O(l) \tag{19}$$

$$2MO \rightarrow 2M + O_2(g) \tag{20}$$

$$MO + OH^- \rightarrow MOOH + e^- \tag{21}$$

$$MOOH + OH^- \rightarrow M + O_2(g) + H_2O + e^- \tag{22}$$

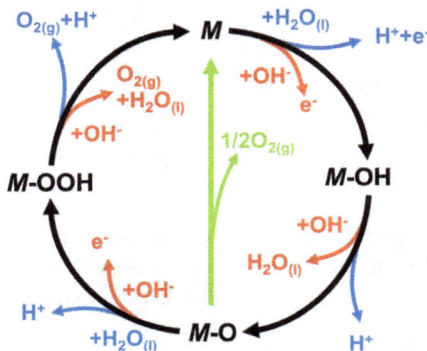

Figure 6. The generally accepted mechanism for OER in acid (blue line), and alkaline (red line) electrolytes [49].

$\triangle G_{O*} - \triangle G_{OH*}$, gives the difference between the energy states of two subsequent intermediates, as the descriptor to reveal the catalytic activity [51]. If the ability of surfaces binding oxygen is too weak, the overpotential would be limited by the oxidation process of OH* because that the intermediates cannot react easily with oxygen. If the binding energy for oxygen absorption is quite strong, the overpotential would be limited by the release of O_2 and the generation of HOO* because the adsorbed products and the intermediates are too stable to either release or reactions with other species [52]. A volcano plot is also obtained via using the $\triangle G_{O*} - \triangle G_{OH*}$ as abscissa, and the current density or overpotential as ordinate.

Similar to HER, the overpotential, exchange current density, and Tafel slope are also key evaluation indexes for OER, and the determination of these indexes are also the same[53]. The faradic efficiency can be determined by quantificationally measuring the amount of generated O_2. Via dividing the experimental generated O_2 by the amount of theoretical generated O_2 during electrolysis, the faradic efficiency can be successfully calculated. Combing these evaluation indexes together, the OER activity of an OER catalyst can be well characterized.

The OER mechanism put forward the requirements for catalysts, including high SSA, high wettability in water, high electrical conductivity, and stability. RuO_2, as one of the typical noble-metal oxides, exhibits excellent catalytic activity toward highly efficient OER in wide-pH range. However, RuO_2 is highly unstable under high anodic potential. IrO_2, another highly active OER catalyst, suffers from a similar problem [54]. Both of them are not practical for large-scale applications owing to their high cost. Thus, substantial efforts have been taken to developing novel OER catalysts with low-cost and investigating their OER mechanism.

3.2 MXene-based catalysts for OER

Although the DFT calculations of MXenes for HER have been widely carried out, the theoretical studies on the OER activities of MXenes have not been carried out. Therefore, MXenes were only considered as potential conductive substrates to modify other active catalysts. To date, various MXene-based hybrid OER catalysts have been designed and reported, including MXene/MOFs, MXenes/LDH, MXene/metal oxides, and MXene/2D materials, etc.

The hybrids of MXene and MOF or MOF-derives have shown their potentials to enhance the OER activities of MXenes or MOFs. Wei Huang *et al.* [55] designed a hybrid OER catalyst of $Ti_3C_2T_x$/2D-cobalt 1,4-benzene-dicarboxylate (CoBDC) via an interdiffusion-assisted method, in which the CoBDC enabled the highly porous architecture and large specific active surface area, the $Ti_3C_2T_x$ nanosheets provided rapid charge/ion transfer and facilitated the accessibility of solution to the active sites (Figure 7a)[55].The obtained$Ti_3C_2T_x$-CoBDCreachedthe current density of 10 mA cm^{-2} at an overpotential of 0.41 V with a Tafel slope of ~48 mV dec^{-1} in KOH media. Furthermore, Jiaguo Yu and Ke Fan *et al.* [56] found that MOF-derived nickel−cobalt sulfides on 2D-MXene were also highly efficient OER catalysts with a small potential of 1.595 V (at 10 mA cm^{-2}) and a suppressed Tafel slope of~58 mV dec^{-1}, as shown in Figure 7b[56].

Hierarchical layered double hydroxide (LDH)/MXenes and LDH-derives/MXene with excellent electrical properties and interfacial junction are potential OER catalysts. Jieshan Qiu *et al.* [58] found that strong interfacial interaction and electronic coupling with prominent charge-transfer between the FeNi-LDH nanosheets and MXenes not only improved the total electrical conductivity and structural stability but also accelerates the redox process of FeNi-LDH for OER. The synthesized FeNi-LDH/Ti_3C_2 delivered a low onset overpotential of 240 mV and a small Tafel slope of 43 mV dec^{-1}in alkaline media. Similarly, Qingyu Yan and co-workers [36] reported a hybrid of $Ni_{1-x}Fe_xPS_3$ and MXene ($Ni_{0.7}Fe_{0.3}PS_3$@MXene) through a facile self-assemble process of LDHon MXene surface and a subsequent solid-state sulfurization reaction. The obtained $Ni_{0.7}Fe_{0.3}PS_3$@MXenenano-hybrid shows a low overpotential of 282 mV and a small Tafel slope of 36.5 mV $dec^{-1}$for OER in KOH solution.

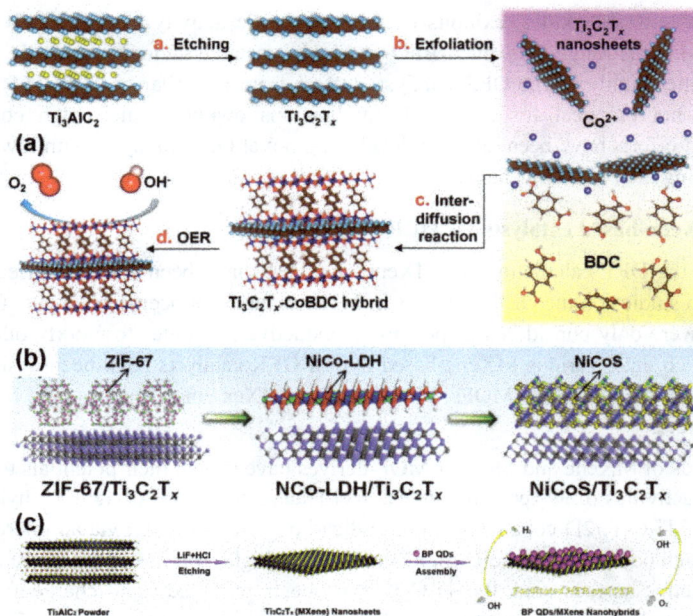

Figure 7. (a) Scheme for the preparation of $Ti_3C_2T_x$-CoBDC [55].(b) Schematics for the synthesis of $NiCoS/Ti_3C_2T_x$ [56]. (c) Schematic illustration for the preparation of BP QDs/MXenes [57].

The transition metal oxides with high activities are potential efficient catalysts for OER. However, their extremely poor electrical conductivity limited their performance. Ke-Ning Sun *et al.* [59] demonstrated a hybrid composed of $Ti_3C_2T_x$ and Co_3O_4 quantum dots for OER and requires a lower overpotential (340 mV) than $Ti_3C_2T_x$ (>650 mV) and Co_3O_4 (423 mV) to reach the current density of 10 mA cm^{-2}[59]. Fan *et al.* [60] reported hierarchical cobalt borates/$Ti_3C_2T_x$ for OER through fast chemical reactions under room temperature, and this obtained novel hybrids showed a low overpotential (250 mV) at 10 mA cm^{-2} and a Tafel slope (~53 mV dec^{-1}). Jianfeng Zhu *et al.* [61] recently reported that S-NiFe$_2$O$_4$/MXene on Ni foam was a highly efficient electrocatalyst for OER. Only 1.50 V was required to reach a relatively high current density of 20 mA cm^{-2} in KOH media. $Ti_3C_2T_x$-metal oxide hydroxides also demonstrated enhanced OER activities. Cui *et al.* [62] designed a $Ti_3C_2T_x$-FeOOH quantum dots structure, and the obtained hybrids delivered a low potential of 1.65 V at 10 mA cm^{-2} and a small Tafel slope of 31.7 mV dec^{-1}.

Other MXene-2D materials, MXene-metal phosphides are also efficient OER catalysts owing to their abundant active sites. Shizhang Qiao *et al.* [63] firstly reported the interacting of carbon nitride (C_3N_4) and Ti_3C_2 could significantly promote the OER performance, in which $Ti-N_x$ acted as active sites. The OER activity of the obtained Ti_3C_2/C_3N_4 is higher than that of IrO_2/C. Black phosphorus (BP) quantum dots (QDs) loaded onMXene 2D-nanosheets through the van der Waals self-assembly, recently was reported by Yi-Tao Liu *et al.* [57] *(Figure 7c).* The 3D-MXene/CoP which is synthesized via Jieshan Qiu *et al.* [38] also showed high OER performance. An overpotential of 298 mV and a Tafel slope of 51 mV $dec^{-1}$was obtained from the optimized 3D-MXene/CoP in alkaline media.

Table 2 summarized the OER performance of reported MXene-based catalysts. Although the hybrid materials of MXenes and other active OER catalysts delivered enhanced activities, the intrinsic characteristics of MXenes for OER have been rarely discussed. In previous reported MXene-based hybrid catalysts, MXenes only acted as conductive substrates. Therefore, more theoretical studies and experimental researches for the OER activities of MXenes are extremely urgent for the development of MXenes in OER fields. The OER processes and mechanism on the surface of MXenes are urgent to be discussed.

Table 2 the MXene-based catalysts for OER

Materials	Solution	Mass loading (mg cm^{-2})	Onset overpotential (V *v.s.* RHE)	Overpotential (V at 10 mA cm^{-2})	Tafel slope (mV dec^{-1})	Ref.
Ti_3C_2/Co-BDC	0.1 M KOH	0.21	0.28	0.41	48.2	[55]
Ti_3C_2/FeNi-LDH	0.1 M KOH	0.2	0.24	0.30	43	[58]
Ti_3C_2/Ni$_{1-x}$Fe$_x$PS$_3$	1 M KOH	0.25	Not given	0.28	36.5	[36]
Ti_3C_2/Co$_3$O$_4$	1 M KOH	1.77	Not given	0.34	64	[59]
Ti_3C_2/NiFe$_2$O$_4$	1 M KOH	Not given	0.08	0.27	46.8	[61]
Ti_3C_2/C$_3$N$_4$	0.1 M KOH	~1.4	0.21	0.42	74.6	[63]
Ti_3C_2/BP	1 M KOH	~0.39	Not given	0.36	64.3	[57]
Ti_3C_2/CoP	1 M KOH	0.2	0.22	0.298	51	[38]

4.   MXene for NRR

4.1   The mechanism of NRR

Ammonia (NH_3) is an extremely important chemical product and has been widely used in agriculture, pharmacology and energy conversion [64]. Remarkably, owing to its high energy density (~2 times larger than that of diesel), environmental benignity (no CO_2 emission), and easy-transportation character, NH_3 is also considered as efficient, clean energy [65]. However, at present, the industrial-scale production of NH_3 is produced through a Haber–Bosch process which was first introduced in 1909. The discovery of the Haber–Bosch process directly lead to a ~4 times increase in agricultural products [66]. However, the Haber–Bosch process requires high temperature (400-500°C), high pressure (200–350 atm), and large amount of H_2 (obtained from $CH_4 + 2H_2 \rightarrow 4H_2 + CO_2$), consuming 1–2% of the world's fossil energy consumption and generating large quantities of CO_2 (nearly 1.5% into the atmosphere) [67]. Therefore, it is of critical significance to develop novel NRR approaches for the production of NH_3.

Naturally, NH_3 can be produced through nitrogenases with an energy transporter of adenosine triphosphate (ATP) under ambient conditions [68]. Inspired by above natural reaction, electrocatalytic and photocatalytic NH_3 generation have been proposed recently, in which the electro(photo-)catalysts the protons and electrons for NRR. Here, we will introduce the mechanism of electro catalytic NRR and their research progress.

The overall NRR reaction can be described as follows [64, 69]:

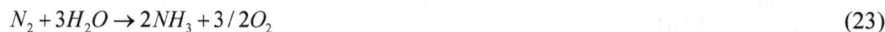

$$N_2 + 3H_2O \rightarrow 2NH_3 + 3/2O_2 \tag{23}$$

In acidic electrolytes:

$$\text{Cathodic reaction:} N_2 + 6H^+ + 6e^- \rightarrow 2NH_3 \tag{24}$$

$$\text{Anodic reaction: } 3H_2O \rightarrow 3/2O_2 + 6H^+ + 6e^- \tag{25}$$

In basic electrolytes:

$$\text{Cathodic reaction:} N_2 + 6H_2O + 6e^- \rightarrow 2NH_3 + 6OH^- \tag{26}$$

Anodic reaction: $6OH^- \rightarrow 3H_2O + 3/2O_2 + 6e^-$ (27)

In general, on the surface of catalysts, NRR processes involve two widely-known reaction mechanisms: i) associative and ii) dissociative mechanisms. However, in the case of the dissociative mechanism, the strong $N\equiv N$ bond is broken before hydrogenation occurs, which is very difficult because $N\equiv N$ bond is one of the strongest in nature (942 kJ mol^{-1}) [70], is not unrealistic. So, here we mainly discuss the associative mechanism, and five possible routes for the NH_3 production are shown in Figure 8 [71]. In path 1, hydrogenation occurs on both sides of N atoms, following by the continuous hydrogenation on other sides, leading to the release of an NH_3 molecule. After that, the continuing hydrogenation process happens on the remaining NH_2 to produce NH_3. In path 2, the N_2 molecule binds to the active surface of the catalyst, and one N atom firstly undergoes the hydrogenation, following by the hydrogenation of the other one N atom, and produce NH_3. In Path 3, the hydrogen atom absorbed on the N atom which is close to the catalyst, resulting in HN*-NH_2 structure. In path 4, the remaining N atoms (after path 1) on the surface absorb H^+ and electrons, realizing the hydrogenation. Path 5 is a combination of path 3 and path 4 [72].

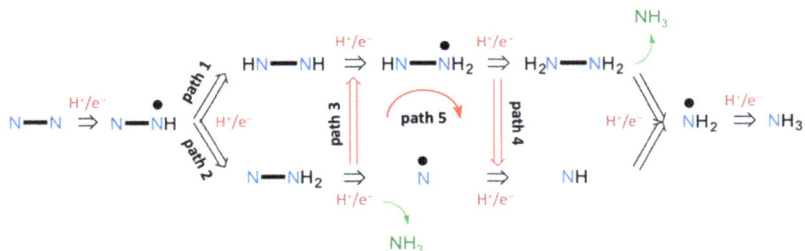

Figure 8. Five proposed routes for the NRR mechanism [71].

The current challenges for NRR can be concluded as follows: 1) The practical mechanism is still not clear and also heavily depends on the catalysts. The theoretical and experimental studies on the reaction processes and immediate seems to be necessary to better understand the reaction mechanism; 2) The major side reaction of the NRR is HER which can be driven under low overpotential, resulting in suppressed faradaic efficiency (FE) of the NRR; 3) the reported material's performance is far away from the requirements for large-scale applications. The yield only reached the level of $\mu g\ h^{-1}$, and the overpotential is up to >0.5 V [66]. Therefore, designing novel NRR catalysts, and studying the NRR mechanism are urgently needed.

4.2   MXene-based catalysts for NRR

DFT calculations indicated that MXene is a promising NRR catalyst. Chenghua Sun *et al.* [71] firstly investigated the d2,d3, and d4 series-MXenes' NRR activities, and concluded that M_3C_2 MXenes are promising catalysts for the capture of N_2, as shown in Figure 9[71]. Among all M_3C_2 MXene materials, V_3C_2 delivered the lowest activation barrier (0.64 eV vs. SHE) in the first step which is the rate-determining step for NRR. This work brought brand-new insights for both MXenes and NRR. Hui Pan *et al.* [73] further studied the NRR mechanism of M_2X MXenes. The results suggested that reaction energy for NRR strongly lies on the charge transfer ability: 1) if N atoms gain electrons during MRR, the total reaction would be exothermic; (2) if N atoms lose electrons during NRR, the total reaction would be endothermic. They also pointed out that two kinds of MXenes: Mo_2C and W_2C, are highly active catalysts among all M_2X MXene materials owing to high exothermic reactions and fast charge transfer characters. Recently, Wang et al. found that 2D-double transition metal carbides can further promote the NRR activities [74]. The overpotential of Mo_2TiC_2 for NRR was reduced to 0.26 V, which is much lower than previously reported MXene NRR catalysts. Remarkably, the potential for HER reached up to 0.74 V, indicating that the competing H_2 evolution was suppressed.

The experimental works further demonstrated that MXenes are highly efficient catalysts for NRR. Wang *et al.* [75] realized highly efficient electrochemical production of NH_3 via taking MXenes as active centers. MXene nanosheet was supported on the surface of a metal substrate with weak hydrogen evolution ability. This structure not only inhibited the competitive HER but also promoted the NRR activity. Diffusion adsorption effectively improves the NRR activity and the selectivity of the MXene catalyst and realized the electrochemical nitrogen fixation under ambient temperature and pressure. According to the theoretical calculation results, the Ti atoms at the edge of the MXene sheets can be used as active sites to preferentially adsorb N_2, thereby performing NRR in an aqueous solution.

MXenes with high electrical conductivities are also ideal substrates to support other catalysts. YupingWu and Yuanhong Xu *et al.* [76] reported a Ti_3C_2/oxygen vacancy-rich TiO_2 catalyst via one-step ethanol thermal treatment for NRR. The Ti_3C_2 was used as a conductive substrate, and the oxygen vacancy-rich TiO_2 provide active sites for NRR. A high NH_3 yield of 32.17 $\mu g\ h^{-1}\ mg^{-1}$was achieved under a potential of -0.55 V. The Faradaic efficiency under a potential of -0.45 V was up to 16.07 %. The DFT calculations also suggest that the composite of TiO_2/Ti_3C_2 has the lowest energy barrier for NRR (0.4 eV).

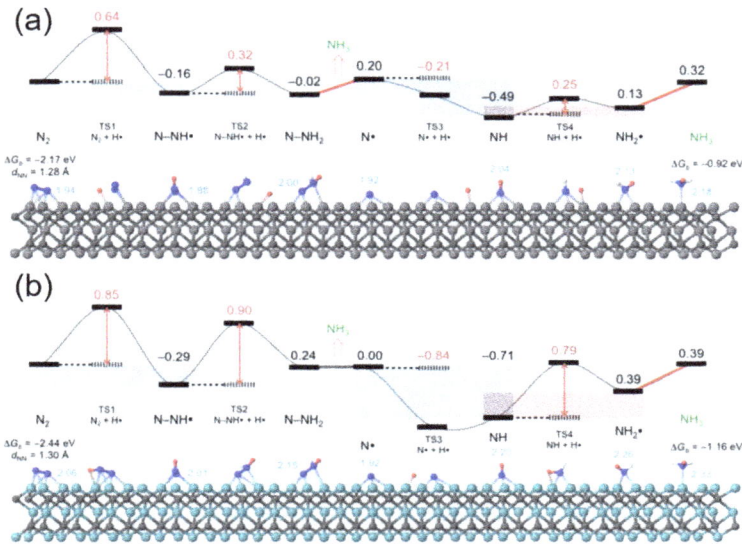

Figure 9. The minimum energy path for the production of NH_3 via NRR for two typical MXenes: V_3C_2 (a) and Nb_3C_2 (b). Gibbs free reaction energies (black, eV) and activation energies (red, eV) are shown here [71].

Although MXenes have shown their great potentials in NRR area, the researches about MXene-based NRR catalysts are still very rare to date owing to the complex processes and strong $N{\equiv}N$ bond. With the growing attentions have been attracted and more efforts have been carried out, the future of NRR and MXene-based catalysts are promising. We believe that the NRR will be a rising star in catalytic fields and attract attentions widely in the next ten years.

Conclusion and outlook

MXene, as a large family of 2D metal carbides/nitrides, which is derived from the selectively etching of "A" phases from MAX phases, show great potential as electrocatalysts. In this chapter, we summarized the mechanism of HER, OER, as well as NRR, and introduced the recent research progress of the structural design of MXene

MXenes: Fundamentals and Applications   Materials Research Forum LLC
Materials Research Foundations 51 (2019) 74-104   doi: https://doi.org/10.21741/9781644900253-4

based catalysts. The advantages of MXenes for electrocatalysis can be concluded as follows:

1) The large MXene family contains ~30-60 kinds of 2D metal carbides or nitrides, suggesting that the electronic structures and properties can be modified via using various metals or using different atom stacking structures (e.g., M_2C, M_3C_2, M_4C_3).

2) The large amounts of functional groups anchored on the surface of MXenes make them easily modify. As discussed above, different termination atoms usually result in different properties. Therefore, applying different termination atoms on the surface of MXenes or doping via various heteroatoms can easily adjust the total electronic structures as well as active sites, and enhance the overall electrocatalytic performance.

3) Most of the MXenes possess high electrical conductivities, which can be utilized as conductive substrates to enhance other active materials' activities.

4) The unique 2D structures of MXenes provide large SSA as well as fast electrochemical reaction surfaces. The ordered 2D channels for ion transport also can promote the mass transfer as well as the electrochemical activities.

However, challenges for the applications of MXenes in electrocatalysis still exist. First, the studies on the MXenes' electrocatalytic performance are just beginning. The mechanism for the electrocatalytic processes is still not clear. The combination of experimental as well as theoretical studies seems to be necessary to figure out the interior mechanism. The kinetics in atomistic to molecular-level remain poorly understood. Second, the advantages of MXenes haven't been fully utilized. Many types of research only take MXenes as conductive materials to enhance the electrical conductivities in their reported hybrid catalysts. However, the electrical conductivities of MXenes are not the highest among all conductive materials, such as graphene, carbon nanotubes (CNTs), etc. [77]. Therefore, how to make full use of the advantages is still a challenge. The rational design of MXene based catalysts is urgent to be studied. Third, although MXene is a large family of 2D metal carbides and nitrides, only ~10-20 kinds of materials among them have been successfully synthesized. So, many non-synthesized MXenes' electrocatalytic properties are still unknown. It is promising that more novel efficient catalysts can be synthesized.

As described here, MXenes promise electrocatalysts for both HER, OER, and NRR. Besides of the electrocatalytic methods we introduced in this chapter, MXenes were also widely studied in many other electrocatalytic fields (e.g., CO_2 reduction, O_2 reduction reaction)[78, 79]. To date, many types of research on water splitting have been carried out. However, the works on MXenes for NRR are very rare owing to the complex reaction process, and obstacles in cleaving N_2. In this regard, we encourage more works can be carried out in NRR which is of critical significance for next-generation catalytic technologies. On the whole, the introduction of MXenes in electrocatalysis provides more possibilities to overcome the obstacles in the electrocatalytic area. It is promising that the large-scale applications of electrocatalytic technologies will come true in the next decade owing to the development of nano-science.

References

[1] H. Jin, C. Guo, X. Liu, J. Liu, A. Vasileff, Y. Jiao, Y. Zheng, S.-Z. Qiao, Emerging two-dimensional nanomaterials for electrocatalysis,Chem. Rev. 118 (2018) 6337-6408. https://doi.org/10.1021/acs.chemrev.7b00689

[2] C.C.L. McCrory, S. Jung, I.M. Ferrer, S.M. Chatman, J.C. Peters, T.F. Jaramillo, Benchmarking hydrogen evolving reaction and oxygen evolving reaction electrocatalysts for solar water splitting devices, J. Am. Chem. Soc. 137 (2015) 4347-4357. https://doi.org/10.1021/ja510442p

[3] L. Zhang, Z.-J. Zhao, J. Gong, Nanostructured materials for heterogeneous electrocatalytic co2 reduction and their related reaction mechanisms, Angew. Chem. Int. Ed., 56 (2017) 11326-11353. https://doi.org/10.1002/anie.201612214

[4] I. Roger, M.A. Shipman, M.D. Symes, Earth-abundant catalysts for electrochemical and photoelectrochemical water splitting, Nat. Rev. Chem. 1 (2017) 0003. https://doi.org/10.1038/s41570-016-0003

[5] Y. Xu, M. Kraft, R. Xu, Metal-free carbonaceous electrocatalysts and photocatalysts for water splitting, Chem. Soc. Rev. 45 (2016) 3039-3052. https://doi.org/10.1039/c5cs00729a

[6] B. Anasori, M.R. Lukatskaya, Y. Gogotsi, 2D metal carbides and nitrides (MXenes) for energy storage, Nat. Rev. Mater. 2 (2017) 16098. https://doi.org/10.1038/natrevmats.2016.98

[7] W. Yuan, L. Cheng, Y. An, S. Lv, H. Wu, X. Fan, Y. Zhang, X. Guo, J. Tang, Laminated hybrid junction of sulfur-doped TiO_2 and a carbon substrate derived from Ti_3C_2 MXenes: toward highly visible Light-driven photocatalytic hydrogen evolution, Adv. Sci. 5 (2018) 1700870. https://doi.org/10.1002/advs.201700870

[8] M. Naguib, O. Mashtalir, J. Carle, V. Presser, J. Lu, L. Hultman, Y. Gogotsi, M.W. Barsoum, Two-dimensional transition metal carbides, ACS Nano 6 (2012) 1322-1331. https://doi.org/10.1021/nn204153h

[9] Y. Zhong, X. Xia, F. Shi, J. Zhan, J. Tu, H.J. Fan, Adv. Sci., 3 (2016) 1500286.

[10] C. Hu, L. Dai, Carbon-based metal-free catalysts for electrocatalysis beyond the ORR, Angew. Chem. Int. Ed. 55 (2016) 11736-11758. https://doi.org/10.1002/anie.201509982

[11] M. Wang, Z. Wang, X. Gong, Z. Guo, The intensification technologies to water electrolysis for hydrogen production–a review, Renew. Sust. Energy Rev. 29 (2014) 573-588. https://doi.org/10.1016/j.rser.2013.08.090

[12] M. Zeng, Y. Li, Recent advances in heterogeneous electrocatalysts for the hydrogen evolution reaction, J. Mater. Chem. A 3 (2015) 14942-14962. https://doi.org/10.1039/c5ta02974k

[13] Y. Yan, B. Xia, Z. Xu, X. Wang, Recent development of molybdenum sulfides as advanced electrocatalysts for hydrogen evolution reaction, ACS Catal. 4 (2014) 1693-1705. https://doi.org/10.1021/cs500070x

[14] J. Duan, S. Chen, M. Jaroniec, S.Z. Qiao, Heteroatom-doped graphene-based materials for energy-relevant electrocatalytic processes, ACS Catal. 5 (2015) 5207-5234. https://doi.org/10.1021/acscatal.5b00991

[15] J. Greeley, T.F. Jaramillo, J. Bonde, I. Chorkendorff, J.K. Nørskov, Computational high-throughput screening of electrocatalytic materials for hydrogen evolution, Nat. Mater. 5 (2006) 909-913. https://doi.org/10.1038/nmat1752

[16] G. Zhao, K. Rui, S.X. Dou, W. Sun, Heterostructures for electrochemical hydrogen evolution reaction: a review, Adv. Funct. Mater. 28 (2018) 1803291. https://doi.org/10.1002/adfm.201803291

[17] Z.W. Seh, K.D. Fredrickson, B. Anasori, J. Kibsgaard, A.L. Strickler, M.R. Lukatskaya, Y. Gogotsi, T.F. Jaramillo, A. Vojvodic, Two-dimensional molybdenum carbide (MXene) as an efficient electrocatalyst for hydrogen evolution, ACS Energy Lett. 1 (2016) 589-594. https://doi.org/10.1021/acsenergylett.6b00247

[18] G. Gao, A.P. O'Mullane, A. Du, 2D MXenes: a new family of promising catalysts for the hydrogen evolution reaction, ACS Catal. 7 (2017) 494-500. https://doi.org/10.1021/acscatal.6b02754

[19] J. Ran, G. Gao, F.-T. Li, T.-Y. Ma, A. Du, S.-Z. Qiao, Ti_3C_2 MXene co-catalyst on metal sulfide photo-absorbers for enhanced visible-light photocatalytic hydrogen production, Nat. Commun. 8 (2017) 13907. https://doi.org/10.1038/ncomms13907

[20] C. Ling, L. Shi, Y. Ouyang, J. Wang, Searching for highly active catalysts for hydrogen evolution reaction based on O-terminated MXenes through a simple descriptor, Chem. Mater. 28 (2016) 9026-9032. https://doi.org/10.1021/acs.chemmater.6b03972

[21] X. Bai, C. Ling, L. Shi, Y. Ouyang, Q. Li, J. Wang, Insight into the catalytic activity of MXenes for hydrogen evolution reaction, Sci. Bulletin 63 (2018) 1397-1403. https://doi.org/10.1016/j.scib.2018.10.006

[22] A.D. Handoko, K.D. Fredrickson, B. Anasori, K.W. Convey, L.R. Johnson, Y. Gogotsi, A. Vojvodic, Z.W. Seh, Tuning the basal plane functionalization of two-dimensional metal carbides (MXenes) To control hydrogen evolution activity, ACS Appl. Energy Mater. 1 (2018) 173-180. https://doi.org/10.1021/acsaem.7b00054

[23] Y.-W. Cheng, J.-H. Dai, Y.-M. Zhang, Y. Song, Two-dimensional, ordered, double transition metal carbides (MXenes): a new family of promising catalysts for the hydrogen evolution reaction, J. Phys. Chem. C 122 (2018) 28113-28122. https://doi.org/10.1021/acs.jpcc.8b08914

[24] C. Ling, L. Shi, Y. Ouyang, Q. Chen, J. Wang, Transition metal-promoted V_2CO_2 (MXenes): anew and highly active catalyst for hydrogen evolution reaction, Adv. Sci. 3 (2016) 1600180. https://doi.org/10.1002/advs.201600180

[25] M.H. Tran, T. Schäfer, A. Shahraei, M. Dürrschnabel, L. Molina-Luna, U.I. Kramm, C.S. Birkel, Adding a new member to the MXene family: synthesis, structure, and electrocatalytic activity for the hydrogen evolution reaction of $V_4C_3T_x$, ACS Appl. Energy Mater. 1 (2018) 3908-3914. https://doi.org/10.1021/acsaem.8b00652

[26] B. Huang, N. Zhou, X. Chen, W.J. Ong, N. Li, Insights into the electrocatalytic hydrogen evolution reaction mechanism on two-dimensional transition-metal carbonitrides (MXene), Chem. Eur. J. 24 (2018) 18479-18486. https://doi.org/10.1002/chem.201804686

[27] Z. Guo, J. Zhou, Z. Sun, New two-dimensional transition metal borides for Li ion batteries and electrocatalysis, J. Mater. Chem. A 5 (2017) 23530-23535. https://doi.org/10.1039/c7ta08665b

[28] W. Yuan, L. Cheng, Y. An, H. Wu, N. Yao, X. Fan, X. Guo, MXene nanofibers as highly active catalysts for hydrogen evolution reaction, ACS Sus. Chem. Eng. 6 (2018) 8976-8982. https://doi.org/10.1021/acssuschemeng.8b01348

[29] X. Yang, N. Gao, S. Zhou, J.J.P.C.C.P. Zhao,MXene nanoribbons as electrocatalysts for the hydrogen evolution reaction with fast kinetics, Phys. Chem. Chem. Phys. 20 (2018) 19390-19397. https://doi.org/10.1039/c8cp02635a

[30] Y. Yoon, A.P. Tiwari, M. Lee, M. Choi, W. Song, J. Im, T. Zyung, H.-K. Jung, S.S. Lee, S. Jeon, K.-S. An, Enhanced electrocatalytic activity by chemical nitridation of two-dimensional titanium carbide MXene for hydrogen evolution, J. Mater. Chem. A 6 (2018) 20869-20877. https://doi.org/10.1039/c8ta08197b

[31] W. Yuan, L. Cheng, H. Wu, Y. Zhang, S. Lv, X. Guo, One-step synthesis of 2D-layered carbon wrapped transition metal nitrides from transition metal carbides (MXenes) for supercapacitors with ultrahigh cycling stability, Chem. Commun. 54 (2018) 2755-2758. https://doi.org/10.1039/c7cc09017j

[32] J. Jia, T. Xiong, L. Zhao, F. Wang, H. Liu, R. Hu, J. Zhou, W. Zhou, S. Chen, Ultrathin N-doped Mo_2C nanosheets with exposed active sites as efficient electrocatalyst for hydrogen evolution reactions, ACS Nano 11 (2017) 12509-12518. https://doi.org/10.1021/acsnano.7b06607

[33] H. Ang, H.T. Tan, Z.M. Luo, Y. Zhang, Y.Y. Guo, G. Guo, H. Zhang, Q. Yan, Hydrophilic nitrogen and sulfur Co-doped molybdenum carbide nanosheets for electrochemical hydrogen evolution, Small 11 (2015) 6278-6284. https://doi.org/10.1002/smll.201502106

[34] S. Zhou, X. Yang, W. Pei, N. Liu, J. Zhao, Heterostructures of MXenes and N-doped graphene as highly active bifunctional electrocatalysts, Nanoscale 10 (2018) 10876-10883. https://doi.org/10.1039/c8nr01090k

[35] D. Geng, X. Zhao, Z. Chen, W. Sun, W. Fu, J. Chen, W. Liu, W. Zhou, K.P. Loh, Direct synthesis of large-area 2D Mo_2C on in situ grown graphene, Adv. Mater. 29 (2017) 1700072. https://doi.org/10.1002/adma.201700072

[36] C.-F. Du, K.N. Dinh, Q. Liang, Y. Zheng, Y. Luo, J. Zhang, Q. Yan, Self-assemble and in situ formation of $Ni_{1-x}Fe_xPS_3$ nanomosaic-decorated MXene hybrids for overall water splitting, Adv. Energy Mater. 8 (2018) 1801127. https://doi.org/10.1002/aenm.201801127

[37] J. Miao, Z. Lang, X. Zhang, W. Kong, O. Peng, Y. Yang, S. Wang, J. Cheng, T. He, A. Amini, Q. Wu, Z. Zheng, Z. Tang, C. Cheng, Polyoxometalate-derived hexagonal molybdenum nitrides (MXenes) supported by boron, nitrogen codoped carbon nanotubes for efficient electrochemical hydrogen evolution from seawater, Adv. Funct. Mater. 29 (2019) 1805893. https://doi.org/10.1002/adfm.201805893

[38] N.H. Attanayake, S.C. Abeyweera, A.C. Thenuwara, B. Anasori, Y. Gogotsi, Y. Sun, D.R. Strongin, Vertically aligned MoS_2 on Ti_3C_2 (MXene) as an improved HER catalyst, J. Mater. Chem. A 6 (2018) 16882-16889. https://doi.org/10.1039/c8ta05033c

[39] X. Wu, Z. Wang, M. Yu, L. Xiu, J. Qiu, Stabilizing the MXenes by carbon nanoplating for developing hierarchical nanohybrids with efficient lithium storage and hydrogen evolution capability, Adv. Mater. 29 (2017) 1607017. https://doi.org/10.1002/adma.201607017

[40] J. Liu, Y. Liu, D. Xu, Y. Zhu, W. Peng, Y. Li, F. Zhang, X. Fan, Hierarchical "nanoroll" like $MoS_2/Ti_3C_2T_x$ hybrid with high electrocatalytic hydrogen evolution

activity, Appl. Catal. B Environ. 241 (2019) 89-94.
https://doi.org/10.1016/j.apcatb.2018.08.083

[41] L. Xiu, Z. Wang, M. Yu, X. Wu, J. Qiu, Aggregation-resistant 3D MXene-based architecture as efficient bifunctional electrocatalyst for overall water splitting, ACS Nano 12 (2018) 8017-8028. https://doi.org/10.1021/acsnano.8b02849

[42] P. Li, J. Zhu, A.D. Handoko, R. Zhang, H. Wang, D. Legut, X. Wen, Z. Fu, Z.W. Seh, Q. Zhang, High-throughput theoretical optimization of the hydrogen evolution reaction on MXenes by transition metal modification, J. Mater. Chem. A 6 (2018) 4271-4278. https://doi.org/10.1039/c8ta00173a

[43] X. Zang, W. Chen, X. Zou, J.N. Hohman, L. Yang, B. Li, M. Wei, C. Zhu, J. Liang, M. Sanghadasa, J. Gu, L. Lin, Self-Assembly of Large-Area 2D Polycrystalline Transition Metal Carbides for Hydrogen Electrocatalysis, Adv. Mater. 30 (2018) 1805188. https://doi.org/10.1002/adma.201805188

[44] J. Zhang, Y. Zhao, X. Guo, C. Chen, C.-L. Dong, R.-S. Liu, C.-P. Han, Y. Li, Y. Gogotsi, G. Wang, Single platinum atoms immobilized on an MXene as an efficient catalyst for the hydrogen evolution reaction, Nat. Catal., 1 (2018) 985-992. https://doi.org/10.1038/s41929-018-0195-1

[45] Y. Yuan, H. Li, L. Wang, L. Zhang, D. Shi, Y. Hong, J. Sun, Achieving highly efficient catalysts for hydrogen evolution reaction by electronic state modification of platinum on versatile $Ti_3C_2T_x$ (MXene), ACS Sus. Chem. Eng. 7 (2019) 4266-4273. https://doi.org/10.1021/acssuschemeng.8b06045

[46] Y. Jiang, X. Wu, Y. Yan, S. Luo, X. Li, J. Huang, H. Zhang, D. Yang, Coupling PtNi ultrathin nanowires with mXenes for boosting electrocatalytic hydrogen evolution in both acidic and alkaline solutions,Small15 (2019) 1805474. https://doi.org/10.1002/smll.201805474

[47] M. Tahir, L. Pan, F. Idrees, X. Zhang, L. Wang, J.-J. Zou, Z.L. Wang, Electrocatalytic oxygen evolution reaction for energy conversion and storage: a comprehensive review, Nano Energy 37 (2017) 136-157. https://doi.org/10.1016/j.nanoen.2017.05.022

[48] F. Lu, M. Zhou, Y. Zhou, X. Zeng, First-row transition metal based catalysts for the oxygen evolution reaction under alkaline conditions: basic principles and recent advances, Small 13 (2017) 1701931. https://doi.org/10.1002/smll.201701931

[49] N.T. Suen, S.F. Hung, Q. Quan, N. Zhang, Y.J. Xu, H.M. Chen, Electrocatalysis for the oxygen evolution reaction: recent development and future perspectives, Chem. Soc. Rev. 46 (2017) 337-365. https://doi.org/10.1039/c6cs00328a

[50] T. Reier, H.N. Nong, D. Teschner, R. Schlögl, P. Strasser, Electrocatalytic oxygen evolution reaction in acidic environments – reaction mechanisms and catalysts, Adv. Energy Mater. 7 (2017) 1601275. https://doi.org/10.1002/aenm.201601275

[51] Z.W. Seh, J. Kibsgaard, C.F. Dickens, I. Chorkendorff, J.K. Norskov, T.F. Jaramillo, Combining theory and experiment in electrocatalysis: Insights into materials design, Science 355 (2017) eaad4998. https://doi.org/10.1126/science.aad4998

[52] Z.-F. Huang, J. Wang, Y. Peng, C.-Y. Jung, A. Fisher, X. Wang, Combining theory and experiment in electrocatalysis: Insights into materials design, Adv. Energy Mater. 7 (2017) 1700544.

[53] C.C.L. McCrory, S. Jung, J.C. Peters, T.F. Jaramillo, Benchmarking heterogeneous electrocatalysts for the oxygen evolution reaction, J. Am. Chem. Soc. 135 (2013) 16977-16987. https://doi.org/10.1021/ja407115p

[54] T. Reier, M. Oezaslan, P. Strasser, Electrocatalytic oxygen evolution reaction (OER) on Ru, Ir, and Pt catalysts: a comparative study of nanoparticles and bulk materials, ACS Catal. 2 (2012) 1765-1772. https://doi.org/10.1021/cs3003098

[55] L. Zhao, B. Dong, S. Li, L. Zhou, L. Lai, Z. Wang, S. Zhao, M. Han, K. Gao, M. Lu, X. Xie, B. Chen, Z. Liu, X. Wang, H. Zhang, H. Li, J. Liu, H. Zhang, X. Huang, W. Huang, Interdiffusion reaction-assisted hybridization of two-dimensional metal–organic frameworks and $Ti_3C_2T_x$ nanosheets for electrocatalytic oxygen evolution, ACS Nano 11 (2017) 5800-5807. https://doi.org/10.1021/acsnano.7b01409

[56] H. Zou, B. He, P. Kuang, J. Yu, K. Fan, Metal–organic framework-derived nickel–cobalt sulfide on ultrathin MXene nanosheets for electrocatalytic Oxygen evolution, ACS Appl. Mater. Interf. 10 (2018) 22311-22319. https://doi.org/10.1021/acsami.8b06272

[57] X.-D. Zhu, Y. Xie, Y.-T. Liu, Exploring the synergy of 2D MXene-supported black phosphorus quantum dots in hydrogen and oxygen evolution reactions, J. Mater. Chem. A 6 (2018) 21255-21260. https://doi.org/10.1039/c8ta08374f

[58] M. Yu, S. Zhou, Z. Wang, J. Zhao, J. Qiu, Boosting electrocatalytic oxygen evolution by synergistically coupling layered double hydroxide with MXene, Nano Energy 44 (2018) 181-190. https://doi.org/10.1016/j.nanoen.2017.12.003

[59] C. Wang, X.-D. Zhu, Y.-C. Mao, F. Wang, X.-T. Gao, S.-Y. Qiu, S.-R. Le, K.-N. Sun, MXene-supported Co_3O_4 quantum dots for superior lithium storage and oxygen evolution activities, Chem. Commun. 55 (2019) 1237-1240. https://doi.org/10.1039/c8cc09699f

[60] J. Liu, T. Chen, P. Juan, W. Peng, Y. Li, F. Zhang, X. Fan, Hierarchical cobalt borate/MXenes hybrid with extraordinary electrocatalytic performance in oxygen

evolution reaction, ChemSusChem 11 (2018) 3758-3765.
https://doi.org/10.1002/cssc.201802098

[61] Y. Tang, C. Yang, Y. Yang, X. Yin, W. Que, J. Zhu, Three dimensional hierarchical network structure of S-NiFe$_2$O$_4$ modified few-layer titanium carbides (MXene) flakes on nickel foam as a high efficient electrocatalyst for oxygen evolution, Electrochim. Acta 296 (2019) 762-770. https://doi.org/10.1016/j.electacta.2018.11.083

[62] N. Li, S. Wei, Y. Xu, J. Liu, J. Wu, G. Jia, X. Cui, Synergetic enhancement of oxygen evolution reaction by Ti$_3$C$_2$T$_x$ nanosheets supported amorphous FeOOH quantum dots, Electrochim. Acta 290 (2018) 364-368.
https://doi.org/10.1016/j.electacta.2018.09.098

[63] T.Y. Ma, J.L. Cao, M. Jaroniec, S.Z. Qiao, Interacting carbon nitride and titanium carbide nanosheets for high-performance oxygen evolution, Angew. Chem. Int. Ed. 55 (2016) 1138-1142. https://doi.org/10.1002/anie.201509758

[64] M. Li, H. Huang, J. Low, C. Gao, R. Long, Y. Xiong, Recent progress on electrocatalyst and photocatalyst design for nitrogen reduction, Small Methods, (2019) 1800388. https://doi.org/10.1002/smtd.201800388

[65] A.R. Singh, B.A. Rohr, J.A. Schwalbe, M. Cargnello, K. Chan, T.F. Jaramillo, I. Chorkendorff, J.K. Nørskov, Electrochemical ammonia synthesis-the selectivity challenge, ACS Catal. 7 (2016) 706-709. https://doi.org/10.1021/acscatal.6b03035

[66] G.-F. Chen, S. Ren, L. Zhang, H. Cheng, Y. Luo, K. Zhu, L.-X. Ding, H. Wang, Advances in electrocatalytic N$_2$ reduction—strategies to tackle the selectivity challenge, Small Methods, (2018) 1800337. https://doi.org/10.1002/smtd.201800337

[67] D. Bao, Q. Zhang, F.L. Meng, H.X. Zhong, M.M. Shi, Y. Zhang, J.M. Yan, Q. Jiang, X.B. Zhang, Electrochemical Reduction of N$_2$ under Ambient Conditions for Artificial N$_2$ Fixation and Renewable Energy Storage Using N$_2$/NH$_3$ Cycle, Adv. Mater. 29 (2017) 1604799. https://doi.org/10.1002/adma.201604799

[68] X. Li, T. Li, Y. Ma, Q. Wei, W. Qiu, H. Guo, X. Shi, P. Zhang, A.M. Asiri, L. Chen, B. Tang, X. Sun, Boosted electrocatalytic N$_2$reduction to NH$_3$by defect-rich MoS$_2$nanoflower, Adv. Energy Mater. 8 (2018) 1801357.
https://doi.org/10.1002/aenm.201801357

[69] N. Cao, G. Zheng, Aqueous electrocatalytic N$_2$ reduction under ambient conditions, Nano Res. 11 (2018) 2992-3008. https://doi.org/10.1007/s12274-018-1987-y

[70] A.J. Medford, M.C. Hatzell, Photon-driven nitrogen fixation: Current progress, thermodynamic considerations, and future outlook, ACS Catal. 7 (2017) 2624-2643. https://doi.org/10.1021/acscatal.7b00439

[71] L.M. Azofra, N. Li, D.R. MacFarlane, C. Sun, Promising prospects for 2D d2–d4 M_3C_2 transition metal carbides (MXenes) in N_2 capture and conversion into ammonia, Energy Environ. Sci. 9 (2016) 2545-2549. https://doi.org/10.1039/c6ee01800a

[72] C. Guo, J. Ran, A. Vasileff, S.-Z. Qiao, Rational design of electrocatalysts and photo (electro) catalysts for nitrogen reduction to ammonia (NH_3) under ambient conditions, Energy Environ. Sci. 11 (2018) 45-56. https://doi.org/10.1039/c7ee02220d

[73] M. Shao, Y. Shao, W. Chen, K.L. Ao, R. Tong, Q. Zhu, I.N. Chan, W.F. Ip, X. Shi, H. Pan, Efficient nitrogen fixation to ammonia on MXenes, Phys. Chem. Chem. Phys. 20 (2018) 14504-14512. https://doi.org/10.1039/c8cp01396a

[74] Y. Gao, Y. Cao, H. Zhuo, X. Sun, Y. Gu, G. Zhuang, S. Deng, X. Zhong, Z. Wei, X. Li, J.-g. Wang, Mo2TiC2 MXene: A Promising Catalyst for Electrocatalytic Ammonia Synthesis, Catal. Today, (2019). https://doi.org/10.1016/j.cattod.2018.12.029

[75] Y. Luo, G.-F. Chen, L. Ding, X. Chen, L.-X. Ding, H. Wang, Efficient electrocatalytic N_2 fixation with MXene under ambient conditions, Joule 3 (2019) 279-289. https://doi.org/10.1016/j.joule.2018.09.011

[76] Y. Fang, Z. Liu, J. Han, Z. Jin, Y. Han, F. Wang, Y. Niu, Y. Wu, Y. Xu, High-performance electrocatalytic conversion of N_2 to NH_3 using oxygen-vacancy-rich TiO_2 in situ grown on $Ti_3C_2T_x$ MXene, Adv. Energy Mater.9 (2019) 1803406. https://doi.org/10.1002/aenm.201803406

[77] W. Yuan, Y. Zhang, L. Cheng, H. Wu, L. Zheng, D. Zhao, The applications of carbon nanotubes and graphene in advanced rechargeable lithium batteries, J. Mater. Chem. A 4 (2016) 8932-8951. https://doi.org/10.1039/c6ta01546h

[78] A.D. Handoko, K.H. Khoo, T.L. Tan, H. Jin, Z.W. Seh, Establishing new scaling relations on two-dimensional MXenes for CO_2 electroreduction, J. Mater. Chem. A 6 (2018) 21885-21890. https://doi.org/10.1039/c8ta06567e

[79] Z. Li, Z. Zhuang, F. Lv, H. Zhu, L. Zhou, M. Luo, J. Zhu, Z. Lang, S. Feng, W. Chen, L. Mai, S. Guo, The marriage of the FeN$_4$ moiety and MXene boosts oxygen reduction catalysis: Fe3d electron delocalization matters, Adv. Mater. 30 (2018) 1803220. https://doi.org/10.1002/adma.201803220

MXenes: Fundamentals and Applications   Materials Research Forum LLC
Materials Research Foundations **51** (2019) 105-136   doi: https://doi.org/10.21741/9781644900253-5

Chapter 5

MXenes Composites

Aqib Muzaffar[1], M. Basheer Ahamed[1,*], Kalim Deshmukh[2]

[1]Department of Physics, B.S. Abdur Rahman Crescent Institute of Science and Technology, Chennai-600048, Tamil Nadu, India

[2]New Technologies- Research Center, University of West Bohemia, Univerzitnı´ 8, 30614, Plzen, Czech Republic

basheerahamed@crescent.edu

Abstract

The chemical transformations of MAX phase compounds especially in the form of chemical exfoliations into new functional and novel two dimensional (2D) nitrides and carbides lead to the formation of so called MXenes. The evolution of these 2D novel materials in the field of material science and technology has brought new opportunities, and their continuous exploitation has opened a new window in the field of electronics. MXenes form a family of layered 2D materials having the combined hydrophilic surfaces with metallic conductivity. The delamination of MXenes produces single–layered nanosheets with a thickness of the order of nanometers with lateral size of the order of micrometers. The higher aspect ratio of these delaminated layers renders MXene promising nanofillers for multifunctional polymeric composites. The addition of other nanofillers to MXene forming hybrid fillers for polymeric composites has been an innovative approach to yield multifunctional materials. This chapter highlights the fundamentals of MXene composites along with their physical and chemical characteristics and potential applications.

Keywords

2D Materials, MXenes, MAX Phases, Polymer Composites, Applications

Contents

1.   Introduction

The exploration of 2D materials initiated during the early 1950s with limited conductivity. However, in recent years, 2D MXenes have been in trend as a smart material exhibiting excellent mechanical and electrical properties [1,2]. During the early 21st century, a new compound based on transition metal carbides was added in the form of 2D MXenes possessing the ability to combine the conductivity of transition metals and the hydrophilic surfaces [3]. MXenes designate the group of compounds produced via etching out the A layer from a layered compound of general formula $M_{n+1}AX_n$ phases which are analogous to graphene [4]. In the formula $M_{n+1}AX_n$ or simply MAX phase, M stands for a transition metal, A comprises of group 13-16 elements, and X or X_n commonly comprises of nitrogen or carbon. The name MXenes originates from MAX phases on the removal of A elements due to their similarity to graphene. The MXene family usually includes Ti_2C, Ti_3C_2, $(V_{0.5},Cr_{0.5})_3C_2$, $(Ti_{0.5},Nb_{0.5})_2C$, Ti_3CN, Ta_4C_3, V_2C, Nb_2C and Nb_4C_3 [5-7]. Currently, more than 70 MAX phases are known, and with the attention diverted in their exploration, more MXenes are expected to surface in the upcoming years [8]. In MXenes the compound surfaces are terminated by OH, H or F group during etching and are termed as $M_{n+1}X_nT_x$ with T denoting the terminating groups. MXenes without delamination exhibit the same multilayered structure as that of exfoliated graphite and as such find applicability in energy storing devices like supercapacitors and lithium ion batteries [9, 10]. The delamination of multilayered MXenes into single or few layered materials leads to enhancement in surface area, and a large quantity of such delaminated layers have been produced via sonication at room temperature [11]. The layers of the delaminated MXenes are of few nanometres thickness with lateral dimensions ranging from 100 nm to few micrometers [12]. The conductive 2D MXenes (excluding graphene based) are very less in number in the form of highly flexible films [13-15]. The trend of producing hybrid MXenes has paved a new way of enhancing conductivities due to the addition of conductive graphene or carbon nanotubes [16].

The low dimensional systems especially MXenes containing transition metals with open d-orbital's display interesting mechanical and other properties due to higher spin orbital coupling and different spin and oxidation states [17]. The low dimensional/2D composites containing transition metals provide an excellent platform for exploration and exploitation their charge, spin, orbital information and behavior, internal degrees of freedom and their device applications [18-20]. Among the low dimensional transition metal composites, MXenes have demonstrated themselves as cutting edge materials enabling new promising technological and scientific perspective [21]. The progress in MXenes has led to the production of high quality and crystalline MXenes using chemical vapour deposition method [22]. Not only that, the MXene family has been extended to double transition metal carbides [23]. With MXenes, there exit a large variety of compositions based on MAX phases leading to the production of a variety of 2D MXenes with extraordinary properties and applicability. For MXenes, theoretically wide range of applications can be predicted like electronics, optical, thermal, thermoelectric, magnetic and sensing, along with experimental applicability like transparent conductors, field effect transistors, hybrid nanocomposites, providing surface for dyes, as fillers in polymer based composites, purifiers, dual responsive surfaces, catalysts and electromagnetic shielding devices, in addition to energy storage devices [24-33]. The large band gap in some MXenes due to d-orbital, low work function, and Schottky barrier junction predicts their insulating capability [34-36]. MXenes in recent times have been attention seekers due to their mechanical and chemical stability resulting from their ceramic nature, existence in various forms, synthesis flexibility, and controllable thickness due to quantum confinement, surface functionalization and demonstration of massless Dirac

MXenes: Fundamentals and Applications   Materials Research Forum LLC
Materials Research Foundations 51 (2019) 105-136   doi: https://doi.org/10.21741/9781644900253-5

dispersions near Fermi level in their band gap [37-40]. The possibilities with MXenes are broad and wide open for exploration.

This chapter briefly describes the significance of MXenes processing, their structural, mechanical, electronic, surface state, transport and optical properties along with their applications in different fields.

2.   Significance of MXenes composites

MXenes especially 2D metal carbides exhibit a unique blend of outstanding metallic conductivity, hydrophilic surfaces, and mechanical properties. These thin layered compounds are incorporated in the polymers to form composites with enhanced flexibility, mechanical strength, and adjustable electronic conductivity. In addition to that, the MXene composites demonstrate excellent volumetric capacitance than the individual components of the composite. Due to their excellent capacitive and mechanical properties, their potential significance is constrained to electrochemical actuators, electromagnetic shielding devices, electrochemical storage and several other applications as well.

3.   MAX phases in MXenes

The family is comprising of layered hexagonal ceramic having chemical compounds having chemical formula $M_{n+1}AX_n$ (n = 1–3) with M, A, and X, representing early transition metal, group 13-16 element, and nitrogen and carbon respectively along with the symmetry group P63/mmc [41]. The emblematic crystal structure of M_2AX MAX phase is shown in Fig.1 [17]. The MAX phase compounds are designated by encouraging ceramic properties like highly stiff structure in addition to favourable metallic properties like better thermal conductivity [43]. Due to combined properties of ceramics and metals, MAX phase compounds have great potential in applications pertaining to corrosion resistance, high wear and tear, lubricants, heat exchangers, nozzles, etc. [42, 43].

The theoretical approach on MAX phase compounds enables studies pertaining to elastic constants of more than 250 M2AC and M2AN compounds with different M and A elements using first principle [41]. Among these theoretically explored MAX phase compounds, many satisfy the mechanical criteria for stability related to hexagonal structures in order $C_{11} > |C_{12}|$, $C_{66} > 0$, and $(C_{11} + C_{12}) C_{33} - 2C_{13}^2 > 0$, where C_{ij} denotes the elastic constant in the form of second order tensor [44]. As such based on their mechanical stability, there is a scope for their experimental exploration. Based on computational studies, the energies formulated for 216 pure M_2AX phase compounds and solid solutions with formulation (MM') (AA') (XX') comprising of 10314 solutions.

MXenes: Fundamentals and Applications   Materials Research Forum LLC
Materials Research Foundations **51** (2019) 105-136 doi: https://doi.org/10.21741/9781644900253-5

Among them, 3140 compounds with addition of 49 experimentally M_2AX phase compounds display the values of formation energies < 30meV per atom and 301 compositions with values of formation < 100 meV per atom based on complimentary synthesis environment [43]. The family of MAX phase compounds has been extended to M_2AlB group of compounds as well. The formation energy (E_f) calculation lies in the order $E_f(M_2AlN)< E_f(M_2AlC)<E_f(M_2AlB)$. The basics pertaining to E_f in MAX phase compounds can be depicted from the hybridization strength between d orbitals of M elements, p orbitals in Al and both s and p orbitals in X elements. In addition to that, the volume expansion energy required for transition metals to host X elements and the valence electrons filling the bonding, antibonding and nonbonding states depict the formation energy in MAX phase compounds [45].

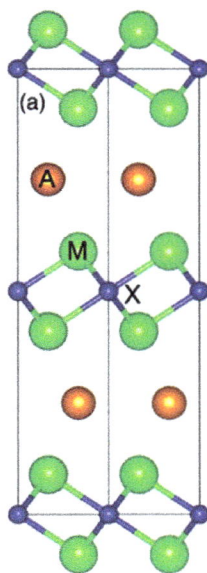

Fig. 1 Emblematic crystal structure of M_2AX MAX phase. Adapted from Ref. [17].

The experimental formation of 2D MXenes consists of displacement of element 'A' in MAX phase on the application of suitable acidic solution like hydrogen fluoride (HF)

[21]. The displacement occurs due to weaker bonding between A and M elements in MAX phases than that between M and X elements. From experimental results, it can be depicted that the presence of saturated OH group, F or O elements at outer layers of MXenes undergo chemical exfoliation by HF acid thereby displacing element A from MAX phase [46, 47]. The exfoliation process has been theoretically reported in Ti_3AlC_2 to Ti_3C_2 using ab-initio molecular dynamic simulations in the presence of HF acid and water. In the presence of HF acid, spontaneous dissociation takes place leading to H or F displacement at the edge of Ti atoms thereby weakening Al-M bonding and widening the interlayer spacing. The creation and widening interlayer spacing on further addition of HF eventually lead to the formation of AlF_3 and H_2 coming out of MAX phase leaving fluorinated MXene behind [48, 49].

4. Processing of MXene composites

The processing of MXene composites has undergone extensive developments in the past decade. Theoretically, various compositions of MXenes have been predicted, and among them, several compositions have been experimentally prepared via either top down or bottom up fabrication approaches. With time these approaches were advanced with the development of various surface modification procedures. The processing of MXenes and its composites follows two general steps which are elaborated in the next sections.

4.1 Synthesis of MXenes

MXenes usually synthesized by exfoliation method follows a classical top down fabrication process which is analogous to the layered structure of precursors of MXenes as that of graphene. The top down fabrication methods are mainly follows cleavage procedure of bulk precursors classified into two categories as etchants and delaminating intercalants [50]. The MXene precursors are generally classified into two types viz; MAX and non-MAX phase precursors. Currently, more than 90 MAX and non-MAX phase precursors are available for MXene preparation. The MAX phase precursors $M_{n+1}ALX_n$ containing different metal compositions are considered to be more efficient than the non-MAX phase precursors based on the target MXenes [51].

Additionally, the MAX phase precursor along with attached Al and ion-attached MAX phase precursors can also be used like Si and Ga following a general formula $M_{n+1}SiX_n$ and $M_{n+1}GaX_n$ respectively [52]. For example, in the MAX phase precursors like Ti_2AlC and Ti_3AlC_2, there exit weak ionic bonds between Ti_3C_2 and Ti_2C via Al or Ga or Si in other MAX phase precursors. A single MXene stack can be achieved using selective etching of ionic bonded layers. The process consists of treating the precursors with suitable etching reagent like HF followed by sonication or shear force treatment to yield a

Materials Research Forum LLC
doi: https://doi.org/10.21741/9781644900253-5

single layer stack of MXene. The non-MAX phase precursors including $Zr_3Al_3C_5$ were reported [53]. The etching process in such cases despite being analogous to Al attached MAX precursors; the layers must be etched from the precursor pre-production phase to yield Zr_3C_2 MXenes. The preparation methods can be classified into two types based on the composition of etchants as HF-etching and non-HF etching processes.

Generally, HF-etching is widely used due to reagent efficiency for MXene precursor's selective etching. The typical HF-etching comprises of high concentrated HF solution reacting with attached ions leading to the displacement and de-stacking of MXenes [51]. Despite being efficient, HF-etching is harsh and harmful, and to avoid these, various substitution methods have been proposed. Among such substitution methods, the approach based on in situ generation of HF is widely acceptable due to acid reaction with fluorides leading to selective etching of attached ions [54]. The other approach uses high temperature heating of molten fluorides to corrode the attached ions [52]. In another approach, powerful probe sonication is used for etching in the presence of tetrabutylammonium hydroxide as intercalant [55]. It is important to mention here that etching conditions vary in MAX phases containing transition metals depending on their particle size, atomic bonding, and structure of the material. The MXenes reported either theoretically or experimentally are illustrated in Fig. 2 [56]. The experimental results reveal that upon the increasing atomic number of early transition metal (M), stronger and longer duration for etching is required [56]. The longer time and stronger etching is attributed to the M-Al bonding and a large number of valence electrons [57]. The etching process in MXenes is kinetically controlled phenomenon and each MXene requires different time for etching to attain a complete conversion. In general, for every MXene synthesis procedure, there are different etching conditions associated with it to yield a different quality of MXenes.

The other essential parameter in the processing of MXenes is the intercalant delamination. Considering MXenes, two general types of intercalants exist viz; organic and metal ion intercalants. The mechanical exfoliations via mechanical forces are used to de-stack MXene layers which are held by strong inter layer interactions than graphite [58]. The use of intercalants before mechanical forces in the form of ultra-fast stirring or high power sonication for exfoliation of MXenes eases the process. The organic intercalants include polar solvents like dimethyl sulfoxide (DMSO), isopropylamine or large organic base molecules like tetrabutyl ammonium hydroxide (TBAOH), tetrapropylammonium hydroxide (TPAOH), choline hydroxide, n-butylamine while the metal ion intercalants include metal hydroxides in aqueous solution or halide salts [50, 54, 55].

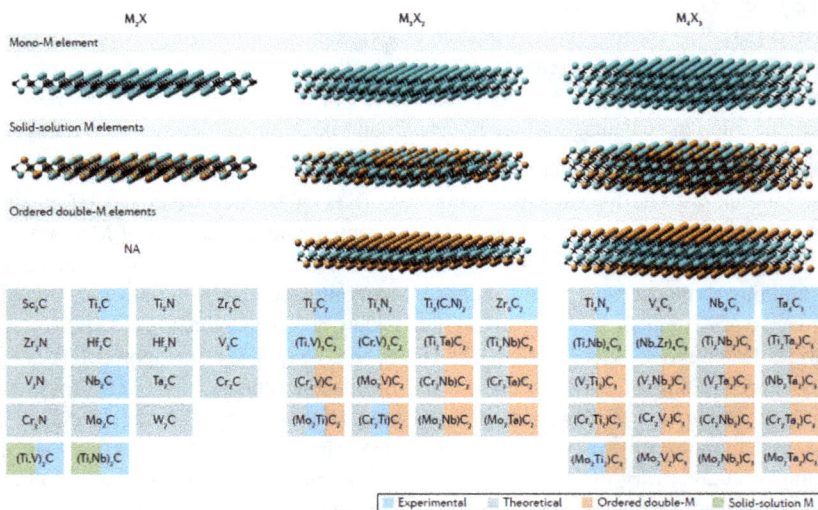

Fig. 2. *Illustration of reported MXenes with three basic formulas M_2X, M_3X_2 and M_4X_3, where M is an early transition metal and X is carbon and/or nitrogen. The central M layers are Ti). Adapted from Ref. [56].*

The synthesis of MXenes has been reported by some bottom-up synthesis methods including chemical vapour deposition to obtain ultrathin (a few nanometres) α-MO_2C orthorhombic 2D crystals with up to 100-μm lateral size [59, 60]. The MXenes comprising of transition metals like tantalum and tungsten were also fabricated by this method in the form of ultrathin crystals. The MXenes synthesized by bottom-up approach exhibits greater lateral size and least defects to enable intrinsic material properties.

4.2   Surface modifications

The MXenes synthesized via various methods are hydrophilic, and as such it is essential to use suitable surface modifications to enable the desired application. The surface modifications of MXenes especially dedicated for biomedical applications result in enhancement of their biocompatibility, loading capacity, and circulation, etc. The surface modification is usually attained using electrostatic attractions or physical absorption [56]. For example, the biocompatibility is induced in some polymers due to surface modifications thereby providing large surface area and biodegradability. The MXene composite composed of soybean phospholipid has been reported as a cost efficient and better competent molecule for effective modification of MXenes [61]. The surface modified MXenes using soybean phospholipid demonstrated stable circulation and exhibited enhanced permeability and retention ability. In another report, the electrostatic attraction was employed for surface modification of MXenes containing positively charged drugs, such as Doxorubicin (DOX) for circulation treatment using negative polymer coating [62]. The synthesized MXenes usually show a negative charge on the surface due to the presence of fluorine or hydroxyl ions, and as such, positively charged molecules can be linked to the surface via electrostatic attraction.

5.   Structural and mechanical properties

The general structure for 2D MXenes is constructed by removal of element 'A' from the bulk MAX phases. The MAX phases exhibit hexagonal symmetry, and the MXenes derives from such phases form hexagonal lattices with same symmetry as shown in Fig. 3 [17]. The 2D M_2X MXene systems as shown in Fig. 3 (a) consist of trilayer sheets having hexagonal unit cells with 'X' sandwiched between two 'M' layers. Generally, the coordination number six forms six coordination bonds with X atoms along with the chemical groups which are attached at the surface [63]. The M_2X surfaces contain two types of hollow sites; one with no X atoms between metal layers and the hollow sites and second with X atoms present at the sites as shown in the lower half of Fig. 3 (a). The positions of terminating groups attached to transition metals via four different configurations are eligible for termination by chemical processes. The first configuration consists of two functional groups at the top of two transition metals while the second one consists of two functional groups positioned at the top of hollow sites denoted by A in the Fig.3. The third configuration comprises of one functional group positioned at the top of hollow site A while the other functional group on the top of another hollow site B as shown in Fig.3. Finally, the fourth configuration comprises two functional groups at the hollow sites as depicted by Fig. 3 (b) and (c), revealing the top and side view of configuration second and fourth respectively. These four configurations form the basis for the functional groups arrangement with free atomic positions to attain the desired stable structure. The stable structure of MXenes is obtained by transformation of the mentioned configurations of functional groups obtained by desired structural optimizations. The relative structure stability mainly depends on the ionic states of the transition metals, capable of providing sufficient electrons towards X and functional groups attached at the surfaces [63].

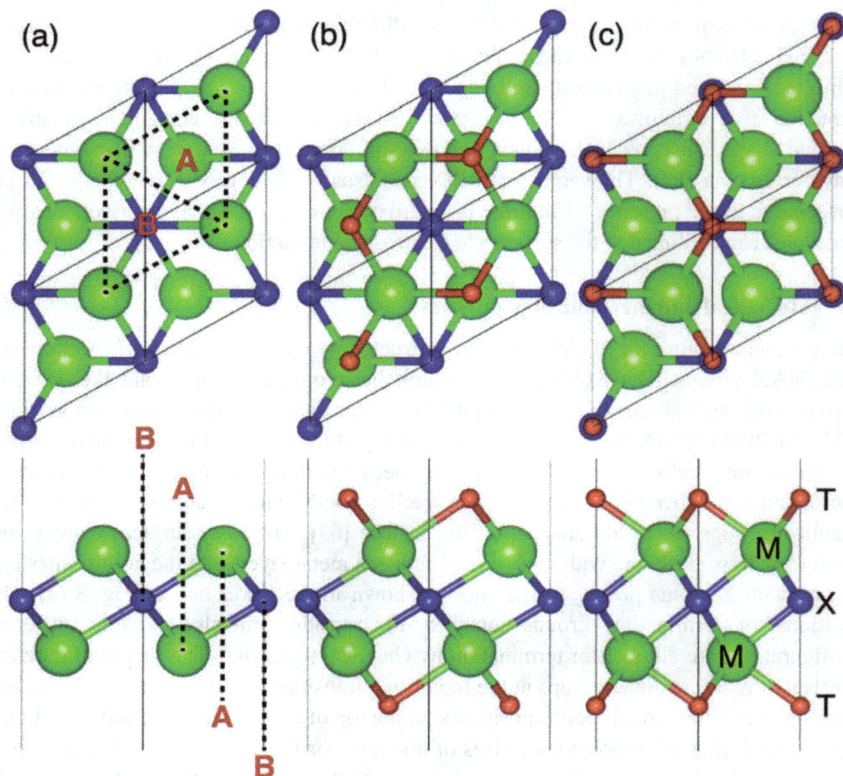

Fig. 3. *Schematic illustrations of M_2X MXenes with top and side views explaining the four general configurations regarding the positioning of functional groups. Adapted from Ref. [17].*

The models for the construction of 2D MXene structures are considered for depicting the stability of functionalized MXene composites and formation of energy evaluations. The evaluations of energy formation from the first principle reveal its large negative value after functionalization of MXene surfaces revealing strong bond formation between functional groups and transition metals. Another aspect about structural properties in MXene composites is the greater thermodynamic stability which favours the formation of MXenes [64]. Experimentally, the MXene composites are not structurally perfect due to the presence of defects at the surface which alters their surface chemistry. The defects are created in composites during etching of surface transition metals in an HF etchant solution. However, the defects can be lowered by varying the etchant concentration thereby enabling the control over surface properties.

The functionalization of MXene surfaces alters the elastic constant of C_{11} of MXenes as revealed by calculations using the first principle [65]. The functionalization of MXenes by incorporation of oxygen show smaller lattice parameters than C_{11} in contrast to functionalization by F or OH [66]. The presence of oxygen induces strong bonding between surface transition metals and oxygen. In addition to functionalization, the mechanical properties also depend on the thickness of MXene composites as reported in 2D $Ti_{n+1}C_n$ (n=1,2 and 3) based composites by molecular simulations. The Young's moduli as calculated from stress-strain curves for the MXene composite were in the range of 500-600 GPa for composites containing MXenes like Ti_2C, TI_3C_2, and Ti_4C_3 with the highest value of modulus for the thinnest one [67]. The tensile strength of Ti_2C and TI_3C_2 based MXene composites also displayed similar behavior under application of high tensile stress where the bond breakage takes place at the outermost regions of the composite layer at the point of higher local stress propagating towards the centre of MXene composite. The crack propagation however occurs via folding and crumpling close to the growing gap. On the other hand in Ti_4C_3 based MXene composite, the crack originates from the centre and propagate from it through the layers and at higher stress levels, the cracks size increases leading to complete fracture of the composite and the formation of composite fragments.

The incorporation of F, O and OH for functionalization in MXene composites also influences the tensile strength and it has been reported theoretically in Ti_2CO_2 based composites which sustain large strain under uniaxial and biaxial tensions [68]. Contrary to that the 2D pristine MXene composites show low tensile strength due to the easy collapse of the surface atomic layer. Therefore, it is essential for such composites to undergo surface modifications to slow the collapse rate a thereby improving mechanical strength and flexibility. The strength enhancement in MXenes via oxygen functionalization occurs due to charge transfer from inner bonds to the outer surface bonds [69]. The functionalization of MXene composites by use of O, F or OH terminations result in weakening of inter layer coupling of MXene elements. Therefore, it is necessary to exfoliate the stacks in composites to form bonds between interlayers to avoid weakening of the composite.

The molecular dynamic and density functional theory demonstrate stiffness and strength of MXene composites as shown in Fig. 4 (a-d) [70] (the experimental testing has been reported on MXene films). The MXene composite composed of polyvinyl alcohol and $Ti_3C_2T_X$ exhibit ~15000 times holding capacity than their weight [70].

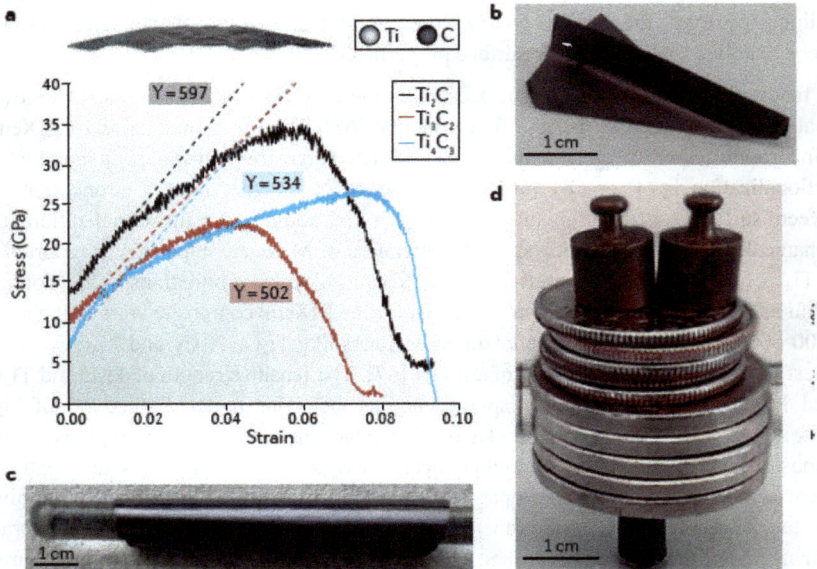

Fig. 4. *Mechanical properties of $Ti_{n+1}C_n$MXenes under molecular dynamics (a) stress-strain curves (b, c) flexibility testing and (d) holding capability test of the composite (90 wt% $Ti_3C_2T_x$–PVA composite). Adapted from Ref. [70].*

6. Electronic properties

The theoretical studies on MXene composites for electronic applications reveal metallic to semiconducting nature of the composite based on the nature of MXene, composite component and surface termination [71-74]. The MXene composites containing heavier transition elements like molybdenum, chromium, and tungsten are predicted to be topological insulators [75-77]. The MXene composites containing Ti_2C, Ti_3C_2, Mo_2C, Mo_2TiC_2 and $Mo_2Ti_2C_3$ mixed terminations have been experimentally demonstrated for electronic applications [78]. The electronic properties of MXenes vary with alterations of outer M layers as shown in Fig. 5 [79]. Despite the metallic nature of $Ti_3C_2T_X$, the MXenes containing Mo exhibit semiconducting properties with positive magneto resistance at 10K as shown in Fig. 5 (a-e). The changes in such MXenes post surface terminations change the transport properties of MXene composites. The electronic properties of the MXenes also depend on the preparation methods. The lower defect concentration and higher flake size yield higher conductivity by mild etching, delamination without being sonicated, better coplanar alignment to ensure good flake contact and finally the drying for intercalant removal [80-82]. For example, the conductivity of $Ti_3C_2T_X$ is increased from 1000 to 6500 Scm^{-1} by mild etching and delamination under vacuum and spin cast films [81]. The MXene composite based on MXene-carbon nanotubes mostly for electrochemical applications utilize electrostatic forces with MXene flakes attaining negative charge with zeta potential between -30 to -80 mV [83]. The self-assembled films of the composite contain positively charged surfactant coated carbon nanotubes.

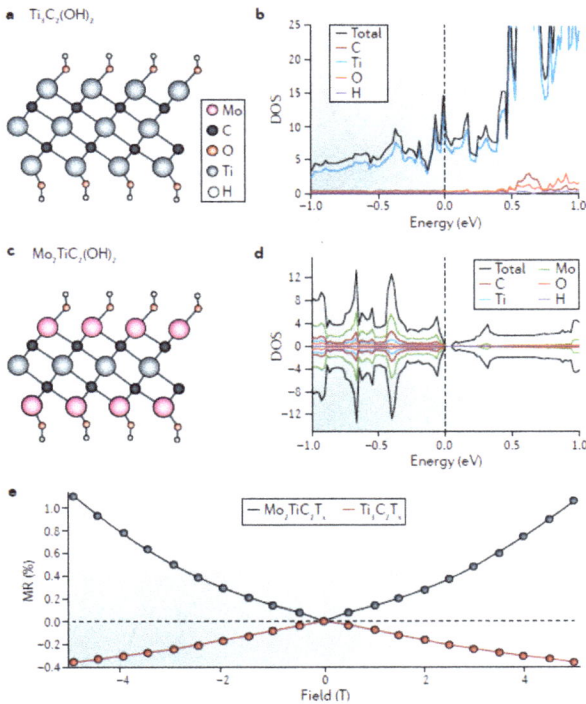

Fig. 5. *Electronic properties of MXenes by varying outer M layers(a, b) Schematic illustration and calculated density of states (DOS) of OH-terminated Ti_3C_2 (c,d) Schematic illustration and calculated DOS of OH-terminated Mo_2TiC_2, respectively. Adapted from Ref. [79].*

7. Surface state properties

In MXene composites, the band structure based on OH group termination reveal some states around G-point in the proximity of Fermi energy without the presence of any elements in the composite [84]. For example, the projected band structure of $Sc_2C(OH)_2$ and $Hf_2C(OH)_2$ are depicted in Fig. 6 [84]. The states in such MXenes are distributed in the vacuum region external to the hydrogen atoms with nearly free electron properties exhibiting parabolic energy dispersion in correspondence to the crystal wave vector.

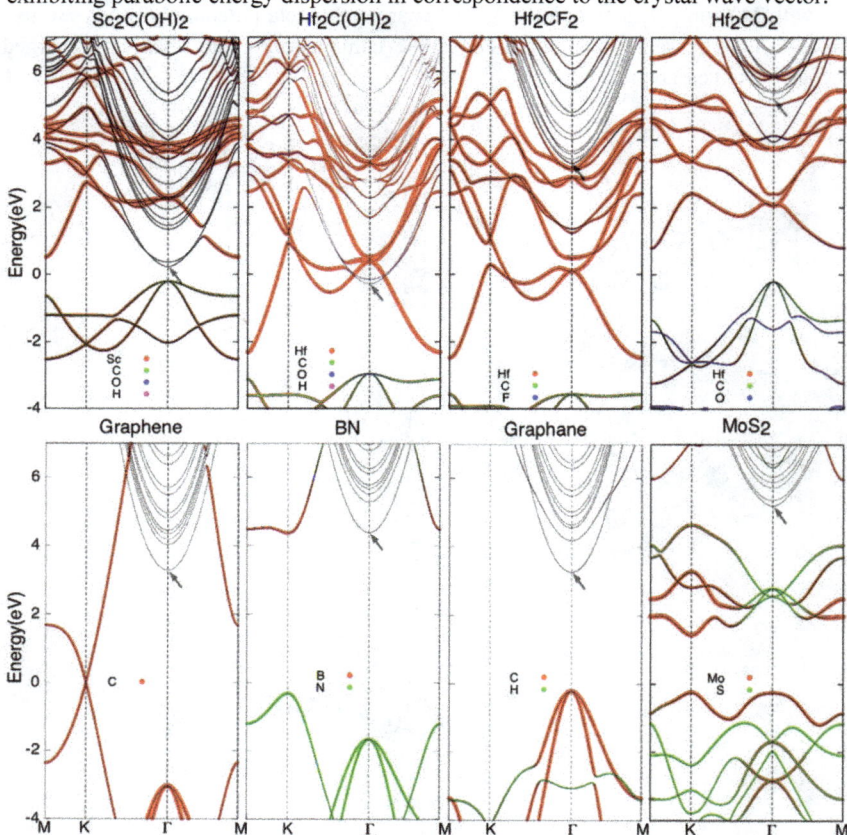

Fig. 6. Proposed band structures for eight diverse systems for each constituent atom. Adapted from Ref. [84].

MXenes: Fundamentals and Applications   Materials Research Forum LLC

Materials Research Foundations **51** (2019) 105-136   doi: https://doi.org/10.21741/9781644900253-5

The electron localization function studies demonstrate the partially occupied nearly free electron states with floating electron gas above hydrogen atoms in MXenes of form $Hf_{n+1}C_n(OH)_2$, where n = 1, 2 and 3 as shown in Fig. 7 [84]. The lowest energy in nearly free electron states is vacant in MXenes like $Sc_2C(OH)_2$, $Ti_2N(OH)_2$ and $V_2C(OH)_2$ and positioned above Fermi energy as shown in Fig. 6. The vacant states in MXenes with OH group termination are easily accessible than the other MXenes and similar trends follow in MXene composites based on the element (O and F) or functional group (OH) terminations. It is noteworthy to mention here that the elemental termination in MXene composites has the nearly free electron states positioned at higher energies thereby restricting the applicability. Contrary to that in OH termination based MXene composites the presence of nearly free electron states in the proximity of Fermi energy is due to the existence of positively charged hydrogen atoms at the surface and independent on the thickness of MXenes. The partially vacant nearly free electron states close to the Fermi energy are sensitive to mechanical forces and processing environment.

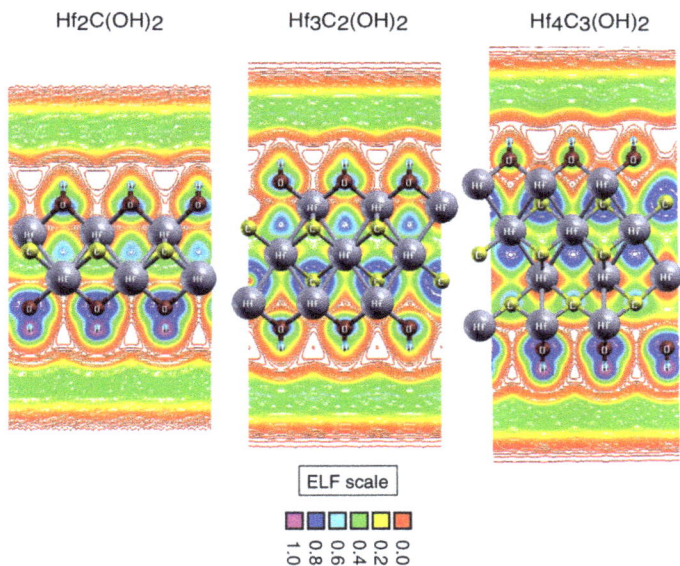

Fig. 7: *Electron localization function for $Hf_{n+1}C_n(OH)_2$, n=1,2 and 3 MXenes varying in thicknesses. Adapted from Ref. [84].*

8.   Transport and optical properties

Considering the transport properties of the MXenes and their composites, only a limited amount of theoretical studies have been put forward. The transport phenomenon calculations are generally attained using non equilibrium greens function and based on that the MXenes and its composites have displayed highly conductive behaviour [85]. By the theoretical analysis on pristine and functionalized Ti_3C_2, the functionalization applied MXene composites produces strong consequences on electron transport. In MXene composite based on Ti_3CF_2, the current values are higher than that in pristine composites at the same bias voltage. The increased value of current is asserted to the smaller variations in electrostatic phenomenon and the emergence of extended electronic states. The electron transport properties in MXene composites are also studied on the basis of partially vacant nearly free electron states responsible for conduction. It is expected that the MXene composites with nearly free electron states show better electron transport properties due to their extension above MXene surfaces thereby conducting electrons via free channels without any alterations from surface vibrations as shown in Fig. 8 [84]. The transport phenomenon based on nearly free electron states behaves as an electron as well as a hole channel under application of lower bias voltage making them ideal for low power electronic or nanoelectronic devices.

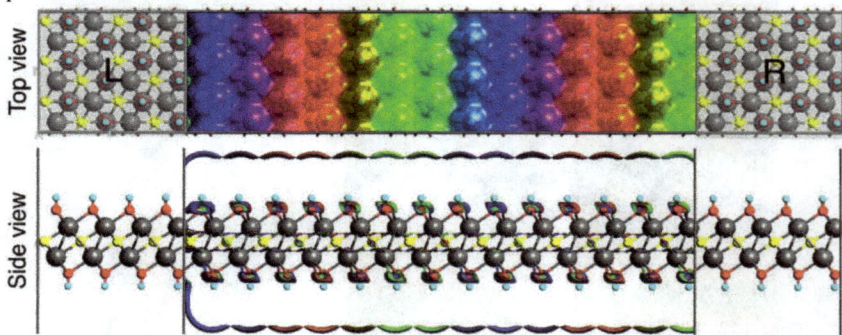

Fig. 8*: Top and side views of one of the eigenvectors of the transmission matrix with NFE characteristics for Hf2C(OH)$_2$. L and R indicate the left and right electrodes, respectively. Adapted from Ref. [84].*

MXenes: Fundamentals and Applications   Materials Research Forum LLC
Materials Research Foundations **51** (2019) 105-136   doi: https://doi.org/10.21741/9781644900253-5

The thin film MXene composites of Ti3C$_2$ are optically transparent with the capability of transmitting 97% of visible light per nanometre thickness [81]. The optical properties of MXene composites can be altered by either by electrochemical or simply by chemical intercalation of cations demonstrating their applicability in optoelectronics and transparent conductive coatings. The evaluations of optical properties like absorption, reflection, scattering, emission, and transmittance in MXene composites are attained using dielectric function tensor as a function of photon wavelength [86]. The theoretical studies on MXene composites confirm their interaction with light. The factors affecting the optical properties of MXene composites include band width and M composition in the visible absorption range. The band gap tuning is attained by surface terminations to enable UV absorption. The MXene composites also demonstrated high photothermal efficiency upto 50%. Also, MXene composites can be proposed as promising substrate candidates for signalling of Raman scattering [87]. The highly metallic conductive MXene composite exhibit and intensive localized surface plasmon resonance effect enabling their applicability as substrates for surface enhanced Raman spectroscopy.

9.   Magnetic properties

The magnetic properties of MXene and their composites are important to determine their biomedical applicability although being least explored. There are only a few theoretical reports predicting the magnetic behavior of MXene composites in correspondence to ferromagnetic behavior or antiferromagnetic characteristics [88]. The magnetic properties of MXene and their composites have been confined to limited experimental studies due to difficulty in termination of free MXene preparation. The magnetic moments in two types of MXenes such as Cr$_2$C and Cr$_2$N with surface terminations were reported without any knowledge of magnetic mechanisms [89]. While considering antiferromagnetic configuration both terminated MXenes were predicted to be antiferromagnetic, but the MXene with oxygen termination remained ferromagnetic with the theoretical results which still lack experimental evidence. The theoretical studies on magnetic properties of MXenes reveal dependence on crystal tension [90]. The ion doping and vacancy importing are expected to control magnetic properties in MXene composites. The ferromagnetic and anti-ferromagnetic behavior of MXene composites is expected with disappearing magnetism during surface terminations [91].

The density functional calculations based on spin polarization predict the nonmagnetic nature for the majority of pristine and functionalized MXene composites due to the existence of strong covalent bonding between the transition metal and X element in addition to the chemical group attachment. However, by application of external strain, the tuning of covalent bonds is attained resulting in the release of 'd' electrons. The electron release leads to the creation of magnetism in a nonmagnetic composite [92]. Some pristine composites are intrinsically magnetic like Ti_2C and Ti_2N exhibiting half metallic ferromagnetism with a magnitude of the magnetic moment as 1.91 and 1 mB per formula unit. In Ti_2C, the monolayer with nearly half metallic ferromagnetism transforming into perfect half metal followed by spin-gapless semiconductor and finally into a metal. The transformation is accompanied by a continuous increase of biaxial strain with monolayers remaining half metallic throughout [93].

Similarly, in MXenes containing V_2C and V_2N, the monolayers are antiferromagnetic and nonmagnetic respectively with biaxial tensile and compressive strain inducing large magnetic moments. The functionalization of such type of MXenes with either OH group or F, V_2 attains characteristics of small gap antiferromagnetic semiconductor [94]. The half metallic characteristic exhibited by many MXene composites makes them promising materials for spintronic devices and applications. Although magnetic properties form an essential parameter to determine the applicability of MXene composites in various fields, but a very limited amount of research is carried out in this direction.

10. Applications of MXenes in different fields

10.1 Low work function emitters

The metallic substrates containing transition metal carbides and nitrides exhibit high stability, melting point, and low work function, due to which they have engrossed significant attention for field emitters. The MXene composites can prove to be ideal candidates for electron emitter applications due to low work function as a consequence of functionalization and tunability of work function by a selection of suitable transition metal and X elements. The surface functionalization during the synthesis of MXene composites leads to variations in electrostatic potential at the surface thereby affecting electronic structures and shifting the Fermi level.

On the other hand, the compositional tunability provides stability and control over thermal, mechanical and chemical properties [95]. The energy difference between the Fermi level and the vacuum level defines the work function of a material. The vacuum level is defined by the electrostatic potential away from the modified surfaces. The work function of MXene and their composites lie in the range of 1 to 7 eV. The work function in MXene composites show a decreasing trend with OH functionalization than the ones functionalized with F or O where work function varies with the type of M elements. The work function in $Ti_{n+1}C_n(OH)_2$/polymer composites depends on the thickness of the layers.

The surface dipole moments help in understanding the work function properties of the MXenes due to the linear correlation of dipole moment with functionalization [96]. The values of dipole moment, either positive or negative corresponds to the increase and decrease in work function respectively due to their linear relation. The dipole moment in MXene composites is controlled by (i) the electron charge redistribution between the adsorbates and the surface (ii) adsorbates surface relaxation and (iii) adsorbates polarity. The analysis of the work function of MXene composites indicates the dependence of work function of functionalized MXene composites (by F or O) on induced dipole moment, surface relaxation and intrinsic dipole moment.

10.2 Catalysts and photocatalysts for hydrogen evolution

In modern times, the use of hydrogen as fuel provides a clean energy source without any pollution [97]. Hydrogen is available in abundance in the form of water and hydrocarbons but its availability in the gas form with high density is available in very less quantity. The evolution of hydrogen is defined as the process of hydrogen production via photocatalysis or electrocatalysis of water or hydrogen containing inorganic molecules. The most efficient catalyst for hydrogen evolution is platinum; however due to its high cost and limited supply, its wide usage is restricted. As catalysts or photocatalysts, MXene composites have attracted much attention as evident from their experimental and theoretical studies. It has been demonstrated theoretically that the MXene composites are conductive under standard pH conditions enabling easy charge transfer during hydrogen evolution.

10.3 Energy conversion for thermoelectric devices

The intrinsic ceramic nature of MXenes and some of its composites make them as suitable for thermo-electronic devices for energy conversion application at reasonably high temperatures. The thermoelectric performance of MXene composites and other devices was measured in terms of ZT dependent on Seebeck coefficient, temperature, electrical and thermal conductivity with electronic and lattice contributions. The ZT varies linearly with power factor and inversely dependent on thermal conductivity. The thermal conductivity is effectively minimized via enhanced phonon scattering in the presence of grain edges and boundaries within the embedded structure [98]. The power factor maximization is a complicated process due to coupling in the electronic structure of the composite, and it varies inversely with Seebeck coefficient and electrical conductivity. It is essential for MXene composites to show a balance between electrical conductivity and Seebeck coefficient at a specific n or p-type carrier concentration to maximize the power component. The MXene composites with metallic and semiconducting behavior show poor and good thermoelectric properties respectively [99].

The thermo electronic property of MXene composite is accredited to unusual band structure near the band edge with merged flat and dispersive bands. This unusual band structure allows a large Seebeck coefficient along with good electrical conductivity even at low carrier concentrations [100].

10.4   Energy storage

The energy storage based on clean energy resources ultimately are hydrogen storage, ion batteries, fuel cells and supercapacitors as the most promising candidates with challenges of their own. However, the major challenge is to find the appropriate materials capable of high storage. The theoretical calculations on pristine MXene composites with MXenes like Sc_2C, V_2C and Ti_2C exhibiting largest surface area among all MXenes. The largest surface area enables high hydrogen storage capacity. The hydrogen storage based on these MXene composites are bound by physical, chemical and Kubas-type interactions with binding energies in the range of 0.5 to 5 eV [101].

Li-ion batteries are capable of producing high gravimetric energy density and high voltage. The higher values of gravimetric energy density containing novel alkali based batteries are the prime focus of the experimental and theoretical explorations [102]. With MXene composites, there arises the capability of intercalation with ions of Li, Na, K, Mg, Al, NH_4, etc. The intercalation of MXenes with ions of these elements has delivered high specific capacitance values up to 350 Fcm^{-1} [103-107]. These overwhelming experimental achievements have triggered an extensive exploration of supercapacitors based on MXene composites. The effects of functionalization on the mobility of various ions and MXenes have also been studied [104-106].

10.5   Biomedical applications

The fascinating physicochemical properties of MXene composites have proven very promising for biomedical applications. The MXene composite has been successfully used in two major biomedical directions like biosensors and bioimaging. MXenes have been broadly used for sensing of gases via alterations of surface terminated groups affecting their properties. The gas sensors using MXene or MXene composites are based mainly on the charge transfer process and conductivity variations during attachment of gas molecules [108]. The small molecular interaction among metal ions and MXenes induces doping effect thereby changing the composite properties. The MXenes composites are also promising candidates for in vivo pH sensors [109]. The protonation and deprotonation of pH sensitive MXene dots decreases the absorption intensity by 10 % with a rise in pH value up to 9. The deprotonation of MXene dot surface results in localization of valence electrons restricting their photo absorption participation. The increase in pH value leads to the transformation of luminescent defects into non luminescent ones thereby increasing non radiative energy losses and decreasing the emission from MXene dots [110].

In bioimaging, MXene composites serve as imaging contrast agents. The current developed imaging technologies based on MXene composite include luminescence imaging as an emerging diagnostic imaging technique capable of penetrating the optical imaging limit and enabling detection of induced pressure waves of laser irradiated tissues [111]. The MXene composite used in imaging for irradiation of tissues with low tissue attenuation coefficient enables real time detection of biological structure and functional information. In addition to the above listed biomedical applications, the MXene composites can also be used in antibacterial activity due to their semiconducting property for killing the bacterial cells. The light based excitation analogous to semiconductors leads to the generation of holes and electrons. Since MXene contain reactive metal pairs for shifting reactive electrons to surrounding cell membranes via photodynamic therapy for killing the cells [112]. MXene composites can also be used for therapeutic applications. MXene composites can serve as carriers of cargo to attain therapeutic effect via generation of reactive oxygen species for killing photodynamic cells, synergistic therapy through cargo loading and cell killing through photothermal effect.

Conclusions

In recent years, the MXene composites have evolved in the field of material science and technology with enormous potential and new opportunities. The MXene composite although being explored highly on a theoretical basis, still lack experimental evidence to support their case for a wide range of applications. This chapter is framed in the context to understand the basics related to MXenes composites from their processing, properties and applications point of view. This new class of materials has great potential in the upcoming years, especially for electronic applications.

References

[1]   K.S. Novoselov, D. Jiang, F. Schedin, T.J. Booth, V.V. Khotkevich, S.V. Morozov, A.K. Geim, Two-dimensional atomic crystals. Proceedings of the National Academy of Sciences, 102 (2005) 10451-10453. https://doi.org/10.1073/pnas.0502848102

[2]   R. Ma, T. Sasaki, Nanosheets of oxides and hydroxides: Ultimate 2D charge-bearing functional crystallites. Adv. Mat., 22 (2010) 5082-5104. https://doi.org/10.1002/adma.201001722

[3]   M. Naguib, O. Mashtalir, J. Carle, V. Presser, J. Lu, L. Hultman, Y. Gogotsi, M.W. Barsoum, Two-dimensional transition metal carbides. ACS Nano, 6 (2012) 1322-1331. https://doi.org/10.1021/nn204153h

[4]   M. Naguib, M. Kurtoglu, V. Presser, J. Lu, J. Niu, M. Heon, L. Hultman, Y. Gogotsi, M.W. Barsoum, Two-dimensional nanocrystals produced by exfoliation of Ti3AlC2. Adv. Mat., 23 (2011), 4248-4253. https://doi.org/10.1002/adma.201102306

[5]   K.S. Novoselov, A.K. Geim, S.V. Morozov, D. Jiang, Y. Zhang, S.V. Dubonos, I.V. Grigorieva, A.A. Firsov, Electric field effect in atomically thin carbon films. Science, 306 (2004) 666-669. https://doi.org/10.1126/science.1102896

[6]   M. Ghidiu, M. Naguib, C. Shi, O. Mashtalir, L.M. Pan, B. Zhang, J. Yang, Y. Gogotsi, S.J. Billinge, M.W. Barsoum, Synthesis and characterization of two-dimensional Nb_4C_3 (MXene). Chem. Comm., 50 (2014) 9517-9520. https://doi.org/10.1039/c4cc03366c

[7]   P. Urbankowski, B. Anasori, T. Er, D. Makaryan, S. Kota, P.L. Walsh, M. Zhao, V.B. Shenoy, M.W. Barsoum, Y. Gogotsi, Synthesis of two-dimensional titanium nitride Ti 4 N 3 (MXene). Nanoscale, 8 (2016) 11385-11391. https://doi.org/10.1039/c6nr02253g

[8]   M.W. Barsoum, M. Radovic, Elastic and mechanical properties of the MAX phases. Ann. Rev. Mat. Res., 41 (2011) 195-227.

[9]   A. Muzaffar, M.B. Ahamed, K. Deshmukh, J. Thirumalai, A review on recent advances in hybrid supercapacitors: Design, fabrication and applications. Renewable and Sustainable Energy Rev., 101 (2019) 123-145. https://doi.org/10.1016/j.rser.2018.10.026

[10]   A. Muzaffar, M.B. Ahamed, Iron molybdate and manganese dioxide microrods as a hybrid structure for high-performance supercapacitor applications. Cer. Int., 45 (2019) 4009-4015. https://doi.org/10.1016/j.ceramint.2018.11.078

[11]   Y. Xie, P.R.C. Kent, Hybrid density functional study of structural and electronic properties of functionalized Ti n+ 1 X n (X= C, N) monolayers. Phy. Rev. B, 87 (2013) 235441.

[12]   D.A. Dikin, S. Stankovich, E.J. Zimney, R.D. Piner G.H. Dommett, G. Evmenenko, S.T. Nguyen, R.S. Ruoff, Preparation and characterization of graphene oxide paper. Nature, 448 (2007) 457. https://doi.org/10.1038/nature06016

[13]   R.K. Joshi, P. Carbone, F.C. Wang, V.G. Kravets, Y. Su, I.V. Grigorieva, H.A. Wu, A.K. Geim, R.R. Nair, Precise and ultrafast molecular sieving through graphene

oxide membranes. Science, 343 (2014) 752-754.
https://doi.org/10.1126/science.1245711

[14]   H. Li, Z. Song, X. Zhang, Y. Huang, S. Li, Y. Mao, H.J. Ploehn, Y. Bao, M. Yu,
Ultrathin, molecular-sieving graphene oxide membranes for selective hydrogen
separation. Science, 342 (2013) 95-98. https://doi.org/10.1126/science.1236686

[15]   R.R. Nair, H.A. Wu, P.N. Jayaram, I.V. Grigorieva, A.K. Geim, Unimpeded
permeation of water through helium-leak–tight graphene-based
membranes. Science, 335 (2012) 442-444. https://doi.org/10.1126/science.1211694

[16]   J.N. Coleman, M. Lotya, A. O'Neill, S.D. Bergin, P.J. King, U. Khan, K. Young,
A. Gaucher, S. De, R.J. Smith, I.V. Shvets, Two-dimensional nanosheets produced by
liquid exfoliation of layered materials. Science, 331 (2011) 568-571.
https://doi.org/10.1126/science.1194975

[17]   M. Khazaei, A. Ranjbar, M. Arai, T. Sasaki, S. Yunoki, Electronic properties and
applications of MXenes: a theoretical review. J. Mat. Chem. C, 5 (2017) 2488-2503.
https://doi.org/10.1039/c7tc00140a

[18]   S.Z. Butler, S.M. Hollen, L. Cao, Y. Cui, J.A. Gupta, H.R. Gutiérrez, T.F. Heinz,
S.S. Hong, J. Huang, A.F. Ismach, E. Johnston-Halperin, Progress, challenges, and
opportunities in two-dimensional materials beyond graphene. ACS Nano, 7 (2013)
2898-2926. https://doi.org/10.1021/nn400280c

[19]   G. Fiori, F. Bonaccorso, G. Iannaccone, T. Palacios, D. Neumaier, A. Seabaugh,
S.K. Banerjee, L. Colombo, Electronics based on two-dimensional materials. Nature
Nanotechnol., 9 (2014)768. https://doi.org/10.1038/nnano.2014.207

[20]   G.R. Bhimanapati, Z. Lin, V. Meunier, Y. Jung, J. Cha, S. Das, D. Xiao, Y. Son,
M.S. Strano, V.R. Cooper, L. Liang, Recent advances in two-dimensional materials
beyond graphene. ACS Nano, 9 (2015) 11509-11539.
https://doi.org/10.1021/acsnano.5b05556

[21]   M. Khazaei, A. Ranjbar, M. Arai, S. Yunoki, Topological insulators in the ordered
double transition metals M 2′ M ″C 2 MXenes (M′= Mo, W; M ″= Ti, Zr, Hf). Phy.
Rev. B, 94 (2016) 125152. https://doi.org/10.1103/physrevb.94.125152

[22]   C. Xu, L. Wang, Z. Liu, L. Chen, J. Guo, N. Kang, X.L. Ma, H.M. Cheng, W.
Ren, 2015. Large-area high-quality 2D ultrathin Mo_2C superconducting
crystals. Nature Mat., 14 (2015) 1135. https://doi.org/10.1038/nmat4374

[23]   B. Anasori, Y. Xie, M. Beidaghi, J. Lu, B.C. Hosler, L. Hultman, P.R. Kent, Y.
Gogotsi, M.W. Barsoum, Two-dimensional, ordered, double transition metals carbides
(MXenes). ACS Nano, 9 (2015), 9507-9516. https://doi.org/10.1021/acsnano.5b03591

Materials Research Forum LLC
doi: https://doi.org/10.21741/9781644900253-5

[24]   J. Halim, M.R. Lukatskaya, K.M. Cook, J. Lu, C.R. Smith, L.Å. Näslund, S.J. May, L. Hultman, Y. Gogotsi, P. Eklund, M.W. Barsoum, Transparent conductive two-dimensional titanium carbide epitaxial thin films. Chem. Mat., 26 (2014) 2374-2381. https://doi.org/10.1021/cm500641a

[25]   Y. Yang, S. Umrao, S. Lai, S. Lee, Large-Area Highly Conductive Transparent Two-Dimensional Ti2CT x Film. J. Phy. Chem. Lett., 8 (2017) 859-865. https://doi.org/10.1021/acs.jpclett.6b03064

[26]   M.R. Lukatskaya, O. Mashtalir, C.E. Ren, Y. Dall'Agnese, P. Rozier, P.L. Taberna, M. Naguib, P. Simon, M.W. Barsoum, Y. Gogotsi, Cation intercalation and high volumetric capacitance of two-dimensional titanium carbide. Science, 341 (2013) 1502-1505. https://doi.org/10.1126/science.1241488

[27]   R.B. Rakhi, B. Ahmed, M.N. Hedhili, D.H. Anjum, H.N. Alshareef, Effect of postetch annealing gas composition on the structural and electrochemical properties of Ti2CT x MXene electrodes for supercapacitor applications. Chem. Mat., 27 (2015) 5314-5323. https://doi.org/10.1021/acs.chemmater.5b01623

[28]   Y. Xie, Y. Dall'Agnese, M. Naguib, Y. Gogotsi, M.W. Barsoum, H.L. Zhuang, P.R. Kent, Prediction and characterization of MXene nanosheet anodes for non-lithium-ion batteries. ACS Nano, 8 (2014) 9606-9615. https://doi.org/10.1021/nn503921j

[29]   F. Shahzad, M. Alhabeb, C.B. Hatter, B. Anasori, S.M. Hong, C.M. Koo, Y. Gogotsi, Electromagnetic interference shielding with 2D transition metal carbides (MXenes). Science, 353 (2016) 1137-1140. https://doi.org/10.1126/science.aag2421

[30]   Q. Peng, J. Guo, Q. Zhang, J. Xiang, B. Liu, A. Zhou, R. Liu, Y. Tian, Unique lead adsorption behavior of activated hydroxyl group in two-dimensional titanium carbide. J. Am. Chem. Soc., 136 (2014) 4113-4116. https://doi.org/10.1021/ja500506k

[31]   G. Fan, X. Li, Y. Ma, Y. Zhang, J. Wu, B. Xu, T. Sun, D. Gao, J. Bi, Magnetic, recyclable Pt y Co 1− y/Ti 3 C 2 X 2 (X= O, F) catalyst: a facile synthesis and enhanced catalytic activity for hydrogen generation from the hydrolysis of ammonia borane. NJC, 41(2017) 2793-2799. https://doi.org/10.1039/c6nj02695h

[32]   F. Liu, A. Zhou, J. Chen, H. Zhang, J. Cao, L. Wang, Q. Hu, Preparation and methane adsorption of two-dimensional carbide Ti 2 C. Adsorption, 22 (2016) 915-922. https://doi.org/10.1007/s10450-016-9795-8

[33]   J. Ran, G. Gao, F.T. Li, T.Y. Ma, A. Du, S.Z. Qiao, Ti 3 C 2 MXene co-catalyst on metal sulfide photo-absorbers for enhanced visible-light photocatalytic hydrogen production. Nature Comm., 8 (2017) 13907. https://doi.org/10.1038/ncomms13907

[34]   M.P. Tran, C. Detrembleur, M. Alexandre, C. Jerome, J.M. Thomassin, The influence of foam morphology of multi-walled carbon nanotubes/poly (methyl methacrylate) nanocomposites on electrical conductivity. Polymer, 54 (2013) 3261-3270. https://doi.org/10.1016/j.polymer.2013.03.053

[35]   M. Khazaei, M. Arai, T. Sasaki, A. Ranjbar, Y. Liang, S. Yunoki, OH-terminated two-dimensional transition metal carbides and nitrides as ultralow work function materials. Phy. Rev. B, 92 (2015) 075411. https://doi.org/10.1103/physrevb.92.075411

[36]   Y. Liu, H. Xiao, W.A. Goddard III, Schottky-barrier-free contacts with two-dimensional semiconductors by surface-engineered MXenes. J. Am. Chem. Soc., 138 (2016) 15853-15856. https://doi.org/10.1021/jacs.6b10834

[37]   H. Weng, A. Ranjbar, Y. Liang, Z. Song, M. Khazaei, S. Yunoki, M. Arai, Y. Kawazoe, Z. Fang, X. Dai, Large-gap two-dimensional topological insulator in oxygen functionalized MXene. Phy. Rev. B, 92 (2015) 075436. https://doi.org/10.1103/physrevb.92.075436

[38]   L. Li, Lattice dynamics and electronic structures of $Ti_3C_2O_2$ and Mo2TiC2O2 (MXenes): The effect of Mo substitution. Com. Mat. Sci., 124 (2016) 8-14. https://doi.org/10.1016/j.commatsci.2016.07.008

[39]   C. Si, K.H. Jin, J. Zhou, Z. Sun, F. Liu, Large-gap quantum spin Hall state in MXenes: d-band topological order in a triangular lattice. Nano Lett., 16 (2016) 6584-6591. https://doi.org/10.1021/acs.nanolett.6b03118

[40]   H. Fashandi, V. Ivády, P. Eklund, A.L. Spetz, M.I. Katsnelson, I.A. Abrikosov, Dirac points with giant spin-orbit splitting in the electronic structure of two-dimensional transition-metal carbides. Phys. Rev. B, 92 (2015) 155142. https://doi.org/10.1103/physrevb.92.155142

[41]   M.F. Cover, O. Warschkow, M.M.M. Bilek, D.R. McKenzie, A comprehensive survey of M_2AX phase elastic properties. J. Phy.: Cond. Matt., 21 (2009) 305403. https://doi.org/10.1088/0953-8984/21/30/305403

[42]   Z.M. Sun, Progress in research and development on MAX phases: a family of layered ternary compounds. Int.Mat. Rev., 56 (2011) 143-166.

[43]   M. Khazaei, M. Arai, T. Sasaki, M. Estili, Y. Sakka, Trends in electronic structures and structural properties of MAX phases: a first-principles study on M_2AlC (M= Sc, Ti, Cr, Zr, Nb, Mo, Hf, or Ta), M2AlN, and hypothetical M2AlB phases. J. Phy.: Cond. Matt., 26 (2014) 505503. https://doi.org/10.1088/0953-8984/26/50/505503

Materials Research Forum LLC
doi: https://doi.org/10.21741/9781644900253-5

[44]   J.F. Nye, Physical properties of crystals: their representation by tensors and matrices. Oxford University Press. (1985).

[45]   X.L. Qi, S.C. Zhang, Topological insulators and superconductors. Rev. Mod. Phy., 83 (2011) 1057.

[46]   M.A. Hope, A.C. Forse, K.J. Griffith, M.R. Lukatskaya, M. Ghidiu, Y. Gogotsi, C.P. Grey, NMR reveals the surface functionalisation of Ti_3C_2 MXene. Phy. Chem.Chem. Phy., 18 (2016) 5099-5102. https://doi.org/10.1039/c6cp00330c

[47]   K.D. Fredrickson, B. Anasori, Z.W. Seh, Y. Gogotsi, A. Vojvodic, Effects of applied potential and water intercalation on the surface chemistry of Ti_2C and Mo_2C MXenes. J. Phy. Chem. C, 120 (2016) 28432-28440. https://doi.org/10.1021/acs.jpcc.6b09109

[48]   P. Srivastava, A. Mishra, H. Mizuseki, K.R. Lee, A.K. Singh, Mechanistic insight into the chemical exfoliation and functionalization of Ti_3C_2 MXene. ACS App. Mat. & Int., 8 (2016) 24256-24264. https://doi.org/10.1021/acsami.6b08413

[49]   A. Mishra, P. Srivastava, H. Mizuseki, K.R. Lee, A.K. Singh, Isolation of pristine MXene from Nb_4AlC_3 MAX phase: a first-principles study. Phy. Chem.Chem. Phy., 18 (2016) 11073-11080. https://doi.org/10.1039/c5cp07609a

[50]   M. Alhabeb, K. Maleski, B. Anasori, P. Lelyukh, L. Clark, S. Sin, Y. Gogotsi, Guidelines for synthesis and processing of two-dimensional titanium carbide (Ti_3C_2Tx MXene). Chem. Mat., 29 (2017) 7633-7644. https://doi.org/10.1021/acs.chemmater.7b02847

[51]   K. Chen, N. Qiu, Q. Deng, M.H. Kang, H. Yang, J.U. Baek, Y.H. Koh, S. Du, Q. Huang, H.E. Kim, Cytocompatibility of Ti_3AlC_2, Ti_3SiC_2, and Ti_2AlN: In Vitro Tests and First-Principles Calculations. ACS BioMat. Sci.& Eng., 3(2017) 2293-2301. https://doi.org/10.1021/acsbiomaterials.7b00432

[52]   M.W. Barsoum, T. El-Raghy, L. Farber, M. Amer, R. Christini, A. Adams, The Topotactic Transformation of Ti_3SiC_2 into a Partially Ordered Electrochem. Soc., 146 (1999) 3919-3923.

[53]   J. Zhou, X. Zha, F.Y. Chen, Q. Ye, P. Eklund, S. Du, Q. Huang, A two-dimensional zirconium carbide by selective etching of Al_3C_3 from nanolaminated Zr3Al3C5. Angewandte Chem. Int. Ed., 55 (2016) 5008-5013. https://doi.org/10.1002/anie.201510432

[54]   M. Ghidiu, J. Halim, S. Kota, D. Bish, Y. Gogotsi, M.W. Barsoum, Ion-exchange and cation solvation reactions in Ti_3C_2 MXene. Chem. Mat., 28 (2016) 3507-3514. https://doi.org/10.1021/acs.chemmater.6b01275

[55]   X. Yu, X. Cai, H. Cui, S.W. Lee, X.F. Yu, B. Liu, Fluorine-free preparation of titanium carbide MXene quantum dots with high near-infrared photothermal performances for cancer therapy. Nanoscale, 9 (2017) 17859-17864. https://doi.org/10.1039/c7nr05997c

[56]   B. Anasori, M.R. Lukatskaya, Y. Gogotsi, 2D metal carbides and nitrides (MXenes) for energy storage. Nature Rev. Mat., 2 (2017) 16098. https://doi.org/10.1038/natrevmats.2016.98

[57]   M. Kurtoglu, M. Naguib, Y. Gogotsi, M.W. Barsoum, First principles study of two-dimensional early transition metal carbides. Mrs Comm., 2 (2012) 133-137. https://doi.org/10.1557/mrc.2012.25

[58]   T. Hu, M. Hu, Z. Li, H. Zhang, C. Zhang, J. Wang, X. Wang, Interlayer coupling in two-dimensional titanium carbide MXenes. Phys. Chem. Chemical Phy., 18 (2016) 20256-20260. https://doi.org/10.1039/c6cp01699e

[59]   Y. Gogotsi, Chemical vapour deposition: transition metal carbides go 2D. Nature Mat., 14 (2015).

[60]   C. Xu, L. Wang, Z. Liu, L. Chen, J. Guo, N. Kang, X.L. Ma, H.M. Cheng, W. Ren, Large-area high-quality 2D ultrathin Mo_2C superconducting crystals. Nature Mat., 14 (2015) 1135. https://doi.org/10.1038/nmat4374

[61]   C. Dai, H. Lin, G. Xu, Z. Liu, R. Wu, Y. Chen, Biocompatible 2D titanium carbide (MXenes) composite nanosheets for pH-responsive MRI-guided tumor hyperthermia. Chem. Mat., 29 (2017) 8637-8652. https://doi.org/10.1021/acs.chemmater.7b02441

[62]   G. Liu, J. Zou, Q. Tang, X. Yang, Y. Zhang, Q. Zhang, W. Huang, P. Chen, J. Shao, X. Dong, Surface modified Ti_3C_2 MXene nanosheets for tumor targeting photothermal/photodynamic/chemo synergistic therapy. ACS App. Mat. & Int., 9 (2017) 40077-40086. https://doi.org/10.1021/acsami.7b13421

[63]   Y. Xie, M. Naguib, V.N. Mochalin, M.W. Barsoum, Y. Gogotsi, X. Yu, K.W. Nam, X.Q. Yang, A.I. Kolesnikov, P.R. Kent, Role of surface structure on Li-ion energy storage capacity of two-dimensional transition-metal carbides. J. Am. Chem. Soc., 136 (2014) 6385-6394. https://doi.org/10.1021/ja501520b

[64]   M. Ashton, K. Mathew, R.G. Hennig, S.B. Sinnott, Predicted surface composition and thermodynamic stability of MXenes in solution. J. Phy. Chem. C, 120 (2016) 3550-3556. https://doi.org/10.1021/acs.jpcc.5b11887

[65]   U. Yorulmaz, A. Özden, N.K. Perkgöz, F. Ay, C. Sevik, Vibrational and mechanical properties of single layer MXene structures: a first-principles

investigation. Nanotechno., 27 (2016) 335702. https://doi.org/10.1088/0957-4484/27/33/335702

[66]   X.H. Zha, K. Luo, Q. Li, Q. Huang, J. He, X. Wen, S. Du, Role of the surface effect on the structural, electronic and mechanical properties of the carbide MXenes. EPL (Europhysics Letters), 111 (2015) 26007. https://doi.org/10.1209/0295-5075/111/26007

[67]   V.N. Borysiuk, V.N. Mochalin, Y. Gogotsi, Molecular dynamic study of the mechanical properties of two-dimensional titanium carbides $Ti_{n+1}Cn$ (MXenes). Nanotechnol., 26 (2015) 265705. https://doi.org/10.1088/0957-4484/26/26/265705

[68]   Z. Guo, J. Zhou, C. Si, Z. Sun, Flexible two-dimensional $Ti_{n+1}C_n$ (n= 1, 2 and 3) and their functionalized MXenes predicted by density functional theories. Phy. Chem. Chem. Phy., 17 (2015) 15348-15354. https://doi.org/10.1039/c5cp00775e

[69]   X. Sang, Y. Xie, M.W. Lin, M. Alhabeb, K.L. Van Aken, Y. Gogotsi, P.R. Kent, K. Xiao, R.R. Unocic, Atomic defects in monolayer titanium carbide (Ti_3C_2T x) MXene. ACS Nano, 10 (2016) 9193-9200. https://doi.org/10.1021/acsnano.6b05240

[70]   Z. Ling, C.E. Ren, M.Q. Zhao, J. Yang, J.M. Giammarco, J. Qiu, M.W. Barsoum, Y. Gogotsi, Flexible and conductive MXene films and nanocomposites with high capacitance. Proceedings of the National Academy of Sciences, 111 (2014) 16676-16681. https://doi.org/10.1073/pnas.1414215111

[71]   X. Zhang, Z. Zhang, Z. Zhou, MXene-based materials for electrochemical energy storage. J. Energy Chem.., 27 (2018) 73-85.

[72]   L. Li, Effects of the interlayer interaction and electric field on the band gap of polar bilayers: A case study of Sc_2CO_2, The J. Phy. Chem. C, 120 (2016) 24857-24865. https://doi.org/10.1021/acs.jpcc.6b08300

[73]   Y. Lee, Y. Hwang, S.B. Cho, Y.C. Chung, Achieving a direct band gap in oxygen functionalized-monolayer scandium carbide by applying an electric field. Phy. Chem. Chem. Phy., 16 (2014) 26273-26278. https://doi.org/10.1039/c4cp03811h

[74]   H. Weng, R. Yu, X. Hu, X. Dai, Z. Fang, Quantum anomalous Hall effect and related topological electronic states. Adv. Phy., 64 (2015) 227-282. https://doi.org/10.1080/00018732.2015.1068524

[75]   B. Silvi, A. Savin, Classification of chemical bonds based on topological analysis of electron localization functions. Nature, 371 (1994) p.683. https://doi.org/10.1038/371683a0

[76]   K.J. Harris, M. Bugnet, M. Naguib, M.W. Barsoum, G.R. Goward, Direct measurement of surface termination groups and their connectivity in the 2D MXene V_2CTx using NMR spectroscopy. The J. Phy. Chem. C, 119 (2015) 13713-13720. https://doi.org/10.1021/acs.jpcc.5b03038

[77]   H. Weng, A. Ranjbar, Y. Liang, Z. Song, M. Khazaei, S. Yunoki, M. Arai, Y. Kawazoe, Z. Fang, X. Dai, Large-gap two-dimensional topological insulator in oxygen functionalized MXene. Phy. Rev.B, 92 (2015) 075436. https://doi.org/10.1103/physrevb.92.075436

[78]   A. Miranda, J. Halim, M.W. Barsoum, A. Lorke, Electronic properties of freestanding $Ti_3C_2T_x$ MXene monolayers. App. Phy. Lett., 108 (2016) 033102. https://doi.org/10.1063/1.4939971

[79]   B. Anasori, C. Shi, E.J. Moon, Y. Xie, C.A. Voigt, P.R. Kent, S.J. May, S.J. Billinge, M.W. Barsoum, Y. Gogotsi, Control of electronic properties of 2D carbides (MXenes) by manipulating their transition metal layers. Nanoscale Hor., 1 (2016) 227-234. https://doi.org/10.1039/c5nh00125k

[80]   A. Muzaffar, K. Muthusamy, M.B. Ahamed, Ferrous nitrate–nickel oxide $(Fe(NO_3)_2$–NiO) nanospheres incorporated with carbon black and polyvinylidenefluoride for supercapacitor applications. J. Electrochem. Energy Con. Stor., 16 (2019) 031008. https://doi.org/10.1115/1.4042727

[81]   A.D. Dillon, M.J. Ghidiu, A.L. Krick, J. Griggs, S.J. May, Y. Gogotsi, M.W. Barsoum, A.T. Fafarman, Highly conductive optical quality solution-processed films of 2D titanium carbide. ADV. Fun. Mat., 26 (2016) 4162-4168. https://doi.org/10.1002/adfm.201600357

[82]   H. Wang, Y. Wu, J. Zhang, G. Li, H. Huang, X. Zhang, Q. Jiang, Enhancement of the electrical properties of MXene Ti_3C_2 nanosheets by post-treatments of alkalization and calcination. Mat. Lett., 160 (2015) 537-540. https://doi.org/10.1016/j.matlet.2015.08.046

[83]   N.C. Osti, M. Naguib, A. Ostadhossein, Y. Xie, P.R. Kent, B. Dyatkin, G. Rother, W.T. Heller, A.C. Van Duin, Y. Gogotsi, E. Mamontov, Effect of metal ion intercalation on the structure of MXene and water dynamics on its internal surfaces. ACS Appl. Mat. & Int., 8 (2016) 8859-8863. https://doi.org/10.1021/acsami.6b01490

[84]   D.J. Singh, Doping-dependent thermopower of PbTe from Boltzmann transport calculations. Phy. Rev. B, 81 (2010)195217. https://doi.org/10.1103/physrevb.81.195217

[85]   G.R. Berdiyorov, Effect of surface functionalization on the electronic transport properties of Ti3C2 MXene. EPL (Europhysics Letters), 111 (2015) 67002. https://doi.org/10.1209/0295-5075/111/67002

[86]   H. Lashgari, M.R. Abolhassani, A. Boochani, S.M. Elahi, J. Khodadadi, Electronic and optical properties of 2D graphene-like compounds titanium carbides and nitrides: DFT calculations. Sol. St. Comm., 195 (2014) 61-69. https://doi.org/10.1016/j.ssc.2014.06.008

[87]   G.R. Berdiyorov, Optical properties of functionalized $Ti_3C_2T_2$ (T= F, O, OH) MXene: First-principles calculations. AIP Adv., 6 (2016) 055105. https://doi.org/10.1063/1.4948799

[88]   H. Kumar, N.C. Frey, L. Dong, B. Anasori, Y. Gogotsi, V.B. Shenoy, Tunable magnetism and transport properties in nitride MXenes. ACS Nano, 11 (2017) 7648-7655. https://doi.org/10.1021/acsnano.7b02578

[89]   A. Chandrasekaran, A. Mishra, A.K. Singh, Ferroelectricity, antiferroelectricity, and ultrathin 2D electron/hole gas in multifunctional monolayer MXene. Nano Lett., 17 (2017) 3290-3296. https://doi.org/10.1021/acs.nanolett.7b01035

[90]   W. Chen, H.F. Li, X. Shi, H. Pan, Tension-tailored electronic and magnetic switching of 2D Ti_2NO_2. The J. Phy. Chem. C, 121 (2017) 25729-25735. https://doi.org/10.1021/acs.jpcc.7b08496

[91]   N.J. Lane, M.W. Barsoum, J.M. Rondinelli, Correlation effects and spin-orbit interactions in two-dimensional hexagonal 5d transition metal carbides, $Ta_{n+1}C_n$ (n= 1, 2, 3). EPL (Europhysics Letters), 101 (2013) 57004. https://doi.org/10.1209/0295-5075/101/57004

[92]   S. Zhao, W. Kang, J. Xue, Manipulation of electronic and magnetic properties of M_2C (M= Hf, Nb, Sc, Ta, Ti, V, Zr) monolayer by applying mechanical strains. App. Phy. Lett., 104 (2014) 133106. https://doi.org/10.1063/1.4870515

[93]   G. Gao, G. Ding, J. Li, K. Yao, M. Wu, M. Qian, Monolayer MXenes: promising half-metals and spin gapless semiconductors. Nanoscale, 8 (2016) 8986-8994. https://doi.org/10.1039/c6nr01333c

[94]   J. Hu, B. Xu, C. Ouyang, S.A. Yang, Y. Yao, Investigations on V_2C and V_2CX_2 (X= F, OH) monolayer as a promising anode material for Li ion batteries from first-principles calculations. The J. Phy. Chem. C, 118 (2014) 24274-24281. https://doi.org/10.1021/jp507336x

[95] L.Y. Gan, Y.J. Zhao, D. Huang, U. Schwingenschlögl, First-principles analysis of MoS_2/Ti_2C and MoS_2/Ti_2CY_2 (Y= F and OH) all-2D semiconductor/metal contacts. Phy. Rev. B, 87 (2013) 245307. https://doi.org/10.1103/physrevb.87.245307

[96] T.C. Leung, C.L. Kao, W.S. Su, Y.J. Feng, C.T. Chan, Relationship between surface dipole, work function and charge transfer: Some exceptions to an established rule. Phy. Rev. B, 68 (2003) 195408. https://doi.org/10.1103/physrevb.68.195408

[97] S. Roldán, C. Blanco, M. Granda, R. Menéndez, R. Santamaría, 2011. Towards a further generation of high-energy carbon-based capacitors by using redox-active electrolytes. Angewandte Chem. Int. Ed., 50 (2011) 1699-1701. https://doi.org/10.1002/ange.201006811

[98] M. Zebarjadi, K. Esfarjani, M.S. Dresselhaus, Z.F. Ren, G. Chen, Perspectives on thermoelectrics: from fundamentals to device applications. Energy & Environ. Sci., 5 (2012) 5147-5162. https://doi.org/10.1039/c1ee02497c

[99] A.N. Gandi, H.N. Alshareef, U. Schwingenschlögl, Thermoelectric performance of the mxenes M_2Co_2 (M= Ti, Zr, or Hf). Chem. Mat., 28 (2016) 1647-1652. https://doi.org/10.1021/acs.chemmater.5b04257

[100] X.H. Zha, J. Yin, Y. Zhou, Q. Huang, K. Luo, J. Lang, J.S. Francisco, J. He, S. Du, Intrinsic structural, electrical, thermal, and mechanical properties of the promising conductor Mo_2C MXene. J. Phy. Chem. C, 120 (2016) 15082-15088. https://doi.org/10.1021/acs.jpcc.6b04192

[101] N.S. Venkataramanan, M. Khazaei, R. Sahara, H. Mizuseki, Y. Kawazoe, First-principles study of hydrogen storage over Ni and Rh doped BN sheets. Chem. Phy., 359 (2009) 173-178. https://doi.org/10.1016/j.chemphys.2009.04.001

[102] K.M. Bui, V.A. Dinh, T. Ohno, Diffusion mechanism of polaron–Li vacancy complex in cathode material Li_2FeSiO_4. App. Phy. Exp., 5 (2012) 125802. https://doi.org/10.1143/apex.5.125802

[103] M.S. Islam, C.A. Fisher, Lithium and sodium battery cathode materials: computational insights into voltage, diffusion and nanostructural properties. Chem. Soc. Rev., 43 (2014) 185-204. https://doi.org/10.1039/c3cs60199d

[104] J.M. Clark, P. Barpanda, A. Yamada, M.S. Islam, Sodium-ion battery cathodes $Na_2FeP_2O_7$ and $Na_2MnP_2O_7$: diffusion behaviour for high rate performance. J. Mat. Chem. A, 2 (2014) 11807-11812. https://doi.org/10.1039/c4ta02383h

[105] P.M. Panchmatia, A.R. Armstrong, P.G. Bruce, M.S. Islam, Lithium-ion diffusion mechanisms in the battery anode material $Li_{1+x}V_{1-x}O_2$. Phy. Chem. Chem. Phy., 16 (2014) 21114-21118. https://doi.org/10.1039/c4cp01640h

[106] Z. Hu, K. Zhang, Z. Zhu, Z. Tao, J. Chen, FeS$_2$ microspheres with an ether-based electrolyte for high-performance rechargeable lithium batteries. J. Mat. Chem. A, 3 (2015) 12898-12904. https://doi.org/10.1039/c5ta02169c

[107] J. Come, M. Naguib, P. Rozier, M.W. Barsoum, Y. Gogotsi, P.L. Taberna, M. Morcrette, P. Simon, A non-aqueous asymmetric cell with a Ti$_2$C-based two-dimensional negative electrode. J. Electrochem. Soc., 159 (2012) A1368-A1373. https://doi.org/10.1149/2.003208jes

[108] X. Zhu, B. Liu, H. Hou, Z. Huang, K.M. Zeinu, L. Huang, X. Yuan, D. Guo, J. Hu, J. Yang, Alkaline intercalation of Ti$_3$C$_2$ MXene for simultaneous electrochemical detection of Cd(II), Pb(II), Cu(II) and Hg(II). Electrochim. Acta, 248 (2017) 46-57. https://doi.org/10.1149/2.003208jes

[109] X. Chen, X. Sun, W. Xu, G. Pan, D. Zhou, J. Zhu, H. Wang, X. Bai, B. Dong, H. Song, Ratiometric photoluminescence sensing based on Ti$_3$C$_2$ MXene quantum dots as an intracellular pH sensor. Nanoscale, 10 (2018) 1111-1118. https://doi.org/10.1039/c7nr06958h

[110] K. Rasool, K.A. Mahmoud, D.J. Johnson, M. Helal, G.R. Berdiyorov, Y. Gogotsi, Efficient antibacterial membrane based on two-dimensional Ti$_3$C$_2$T$_x$ (MXene) nanosheets. Sci. Rep., 7 (2017) 1598. https://doi.org/10.1038/s41598-017-01714-3

[111] K. Huang, Z. Li, J. Lin, G. Han, P. Huang, Two-dimensional transition metal carbides and nitrides (MXenes) for biomedical applications. Chem. Soc. Rev., 47 (2018) 5109-5124. https://doi.org/10.1039/c7cs00838d

[112] X. Han, J. Huang, H. Lin, Z. Wang, P. Li, Y. Chen, 2D ultrathin mxene-based drug-delivery nanoplatform for synergistic photothermal ablation and chemotherapy of cancer. Adv. Healthcare Mat., 7 (2018) 1701394. https://doi.org/10.1002/adhm.201701394

MXenes: Fundamentals and Applications   Materials Research Forum LLC
Materials Research Foundations **51** (2019) 137-174   doi: https://doi.org/10.21741/9781644900253-6

Chapter 6

MXenes for Supercapacitors

Qixun Xia*

School of Materials Science and Engineering, Henan Polytechnic University, Jiaozuo 454000,
China

xqx@hpu.edu.cn

Abstract

As new members of the two-dimensional (2D) material family, MXene have attracted
spectacular attentions due to their potentials for electronic, optoelectronic, catalyze,
biological, gas sensing, and energy storage applications. In these applications,
supercapacitor has been extensively studied, and other applications are also expanding. In
this chapter, recent dramatic developments on MXene-based supercapacitor electrode
materials, such as single/few-layered MXene, element doped MXenes, MXene-based
composites, MXene quantum dots, are highlighted. Furhter, some important progress on
microstructure, electrical properties and supercapacitor applications reviewed. Lastly, a
brief outlook points out future development direction of MXene applications on
supercapacitor devices.

Keywords

MXene, Supercapacitors, Composites, Electrochemical, Two-Dimensional

Contents

1.   Introduction

In recent years, because of the exploitation of fossil fuels, serious environment pollution problems, global warming, and resource shortage, encouraged us to develop advanced energy management devices, such as lithium ion batteries (LIBs), and supercapacitors (SCs). These advanced energy storage devices are required to provide high energy density and high power density [1]. Among of these energy storage device, SCs have attracted attention recently attribute to their reliable safety, high power density, fast charge and discharge rates, excellent cycling performance, and higher environmental suitability than that of LIBs [2, 3].

In a general way, SCs can be classified into two categories, which depends on different electron storage theoretical: electrical double layer capacitors (EDLC, by electrostatic adherence) and Faraday supercapacitors (by rapid surface redox reactions). Electrical double layer capacitors normally utilize carbonaceous materials as the double-layer electrode display long cycling lives but its capacitance is low. Faraday supercapacitors generally use transition metal compound (includes metal oxides, metal hydroxides, and metal sulfide et al.) as electrode materials, demonstrating higher capacitances than those of carbon materials [4].

Among of those SC electrode materials, two-dimensional (2D) materials with special properties and multiformity have captured more and more attention [5]. Graphene as the typical 2D material, has been widely used in the field of SCs with good electrochemical performance due to its high specific surface area and low resistivity [6-8]. However, low volume capacitance limits the practical application of carbon materials. With growing demand for portable batteries, the volumetric capacitance has turn into an significant indicators for evaluating the electrochemical performance of SCs [9]. The specific volumetric electrochemical properties reflects how much energy can be stored per unit volume of electrode materials or packaged device, which manifests more reasonable and accurate parameter for the electrochemical performance appraisal of SC compared with the specific gravimetric capacitance.

Recently, a large but quickly expanding group of newly developed 2D materials is the Mxenes [10]. These materials have been synthesized through selective removal of the A layers from a laminar MAX phase material with the structural formula of $M_{n+1}AX_n$, where M is an early transition metal, A is a IIIA- or IVA-group element, X is carbon and/or nitrogen, and $n = 1$-3 [11, 12]. The MXene family of 2D materials, in addition to outstanding chemical, thermal, and environmental stabilities, has confirmed extraordinary metallic conductivity and hydrophilic [10, 13]. Large area layered structure is an additional asset of this material. MXene electrode materials are being investigated as SCs with superior volumetric capacitance than previously reported carbonaceous materials of nearly or less than 300 F cm^{-3} volumetric capacitance [14-16]. This indicates that MXenes are promising electrode materials in supercapacitors owing to their high parking density [17]. Although several review papers on MXenes have been published [18-23], they generally review all the application on MXene, and a special review on the progress of the application of Mxenes in the adsorptive environmental, hydrogen storage medium, lithium-ion batteries, electronic properties, and energy storage applications, none covers their progress focus on SCs. Herein, the advancement of MXene and their derivatives, encompassing both theoretical and experimental investigations of relevance to their recently progress on prepare approaches, structure characterizations, and electrochemical performances in the SCs area, will be detailed and discussed.

2.   Supercapacitor background

Electrochemical energy storage systems can be divided into three types, which are batteries, fuel cells and SCs, based on different energy conversion mechanisms. The common denominators that the systems share are the separated electron/ion transport and that the energy conversion process happens at the electrode/electrolyte interface. In batteries and fuel cells, energy is stored through redox reactions. Anode, cathode, and electrolyte are the identical module in both battery and fuel cell. Ionic conductivity is offered by electrolyte, which finally allows electric charges transfer to each electrodes. In SCs, electric energy is stored at the interface of electrode/electrolyte, which an electrochemical double layer supercapacitor (EDLC) is formed. Energy storage in the SCs consists of electrostatic storage from separation of charge in a double layer at the electrode/electrolyte interface, and Faradic electrochemical storage from redox reactions. An ideal energy storage system compromises high energy capacity and high rate of energy conversion. Fuel cells have the highest specific energy mainly because of the constant chemical energy source from outside. In addition, battery has a lower energy capacity compared to fuel cells. Capacitors can be considered as high power systems

MXenes: Fundamentals and Applications   Materials Research Forum LLC
Materials Research Foundations **51** (2019) 137-174   doi: https://doi.org/10.21741/9781644900253-6

based on the charge/discharge mechanism. It has several orders of magnitude higher power density than that of battery.

SC was demonstrated and patented by General Electric in 1957 [24]. Base on their charge storage mechanisms, electrochemical capacitors can be divided into two categories, EDLCs and pseudocapacitors. EDLCs normally have good cycle lifetime and high maximum power density. However, EDLC do not have high enough specific capacitance or energy density. Pseudocapacitors, the capacity is due to redox reactions, the redox Pseudocapacitors shows highly reversible Faradic-type charge transfer and the resulting capacitance is not electrostatic adsorption in origin and therefore called the pseudocapacitors, provide higher energy densities than EDLCs.

Capacitance is the important parameter for capacitor, which is defined as the ratio of the charge quantity to the applied potential:

$$C = Q/V \tag{1}$$

In some cases, the capacitance may change with the applied potential. Hence, the capacitance can be defined as the derivative of the charge and applied voltage:

$$C = dQ/dV \tag{2}$$

Generally, a capacitor consists of two parallel plates and separated by a separator, which is dielectric with permittivity ε, d is the thickness of separator; A is the surface area of parallel plates. So, the capacitance of capacitor is:

$$C = \varepsilon A/d \tag{3}$$

Strategies for increasing capacitance can be summarized as follows: 1) use high dielectric constant materials as the separators; 2) shorten the distance between the negative electrodes and positive electrodes, 3) development of new electrode materials with high specific surface area, high theoretical capacitance, and excellent chemical stability. As reported [6, 25, 26], carbon materials, such as graphene, carbon nanotube (CNT), activated carbon et al. with excellent conductivity, good chemical and thermal stability, and high surface area, it is widely used as electrode material for capacitors.

To evaluate the performance of the energy storage device, the performance energy density and power density are the two all-important parameters. The energy density of a capacitor is:

$$E = 0.5CV^2 = Q^2/2C \tag{4}$$

Where E is energy density (kW kg^{-1}), C is the capacitance (F g^{-1}), V is the potential window (V), Q is charge quantity (C) of capacitor. For the power density P, the time $2t$ is the discharge time in the charge/discharge process.

$$C = 2E/(2t) \tag{5}$$

Generally, traditional electrostatic capacitors with gratifyingly power density higher than 10000 W/kg, but energy density very low, lower than 0.1 Wh kg^{-1}. Compared with other energy storage devices, electrostatic capacitors shows vary fast charge/discharged rates, but the amount of energy stored is very limited. However, the supercapacitor using new materials, which shows high specific surface area, high theoretical capacitance, and excellent chemical stability, as the electrode materials of SCs. Compare with the traditional electrostatic capacitors, SCs demonstrating much higher energy density.

The electrochemical properties of SCs were performed by a three-electrode electrochemical system. Cyclic voltammetry (CV) and galvanostatic charge/discharge (GCD) measurements were tested by electrochemical workstation. The specific capacitance (C_s) of the electrode calculated from GCD curves [27]:

$$Cs = (I \cdot \Delta t)/(m \cdot \Delta V) \tag{6}$$

where I (mA) stands for the applied current, and m (mg) is the loading mass of active materials, ΔV (V) is the potential window, and Δt (s) is the discharge time.

The energy density E (Wh kg^{-1}) and the power density P (W kg^{-1}) of the SC devices were derived from the GCD curves according to equations (3) and (4) [28],

$$E = (C_{device} \times \Delta V^2)/7.2 \tag{7}$$

$$P = E/\Delta t \tag{8}$$

where C_{device} (F g^{-1}) is the specific capacitance, E (Wh kg^{-1}) is the energy density, ΔV (V) is the potential window, P (W kg^{-1}) is the power density, and Δt (s) is the discharge time of SC device.

The volumetric energy density (Wh cm^{-3}) and volumetric power density (W cm^{-3}) values of the SCs, based on the GCD curves, were calculated from equations given bellow [28, 29]:

$$E = (I \rho \int V \, dt) / 3.6m \tag{9}$$

$$P = E/\Delta t, \tag{10}$$

where E (Wh cm^{-3}) is the volumetric energy density, I is the applied current, $\int V dt$ is the galvanostatic discharge current area, m is the total mass of active materials, ρ is the mass density of total active materials (include positive and negative electrode materials), P (W cm^{-3}) is the volumetric power density, and Δt (s) is the discharge time of the SC devices.

3. Synthesis approaches

3.1 MXene

MXenes are obtained by selective removal of the A atomic layers from their laminar MAX phases precursors (Figure 1a, b), which pertain to a very large group (more than 70 kinds) of ternary or polynary (multiple solid solutions) carbides and/or nitrides, approximately 30 different MXenes have been synthesized. It is hard to separate the M-X layers and prepare MXenes using a mechanical method to shear MAX phases due to the metallic M-A bond. In 2011, M. Naguib *et al.* [30] successfully synthesised $Ti_3C_2T_x$ MXene for the first time by a HF etching mehod. Powder of Ti_3AlC_2 MAX phase, which is the precursor of $Ti_3C_2T_x$ MXene was synthesized by following steps: 1) ball-milling Ti_2AlC and TiC powders as 1:1 molar ratio for 24 h; 2) the Ti_3AlC_2 MAX phase was obtained by keeping the temperature at 1350 °C in argon atmosphere for 2 h; 3) 10 g of Ti_3AlC_2 MAX phase powders are then immersed in 100 mL hydrofluoric acid solution (concentrated, 50%) at room temperature for 2 h. 4) the suspension was cleaned with DI water for several times and then dried to obtain $Ti_3C_2T_x$ MXene powder. The reaction equation of this exfoliation process can be conclude as follows [31]:

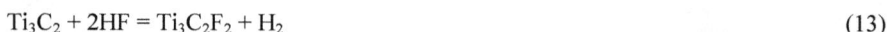

$$Ti_3AlC_2 + 3HF = AlF_3 + 3/2H_2 + Ti_3C_2 \tag{11}$$

$$Ti_3C_2 + 2H_2O = Ti_3C_2(OH)_2 + H_2 \tag{12}$$

$$Ti_3C_2 + 2HF = Ti_3C_2F_2 + H_2 \tag{13}$$

Reaction (1) is basic reactions in the exfoliation process, and Reaction (2) and (3) are the formation process of –OH and/or –F terminations in the exfoliation process.

For the synthesis of the $Ti_3C_2T_x$ MXenes, Kwang Ho Kim's group [32] reported a new method to prepare the precursor of MXene *i.e.* Ti_3AlC_2 MAX phase at a lower calcination temperature (1200 °C). In this work, Ti, Al, and graphite powder were mixed as 3:1:2 molar ratio under ball milling for 12 h. Then the Ti_3AlC_2 MAX, as the precursor of the MXene $Ti_3C_2T_x$, was prepared by the following chemical reaction:

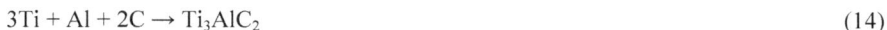

$$3Ti + Al + 2C \rightarrow Ti_3AlC_2 \tag{14}$$

The mixture was put in a graphite mold, then heated at 1200 °C for 10 min under a 30 MPa pressure via a spark plasma sintering (SPS) technology. The Ti_3AlC_2 powder was prepared by ball-milling route. The obtained Ti_3AlC_2 powder was immersed in hydrofluoric acid solution (concentrated, 50%) for 24 h. then washed with DI water many times to remove the hydrofluoric acid. $Ti_3C_2T_x$ MXene was further exfoliated by the insertion of small organic molecules (dimethyl sulfoxide, DMSO) under sonication treatment.

In 2012, Yury Gogotsi's group [10] used the same HF exfoliation method to etch other MAX phases, such as Ti_2AlC, Ta_4AlC_3, $(Ti_{0.5},Nb_{0.5})_2AlC$, $(V_{0.5},Cr_{0.5})_3AlC_2$ and Ti_3AlCN and successfully prepared 2D transition metal carbides or carbonitrides of Ti_2C, Ta_4C_3, $(Ti_{0.5},Nb_{0.5})_2C$, $(V_{0.5},Cr_{0.5})_3C_2$ and Ti_3CN Mxene. In 2013, Nb_2AlC and V_2AlC were etched in a similar way to obtain Nb_2C and V_2C two-dimensional materials [33]. Whereafter, a ordered solid solution MAX phases come to light. Michel W. Barsoum and Yury Gogotsi *et al.* [34] used first-principles calculations based on density functional theory (DFT) to find that there were wxpected to exist of two new types of 2D MXenes, $M'_2M''C_2$ and $M'_2M''_2C_3$, where M' and M" on behalf of two different early transition metals. Then, they synthesized $Mo_2Ti_2C_3T_x$ double transition metals carbides Mxenes (Figure 1c-i) [34]. In this work, the Mo, Ti, Al and graphite powders with 2:2:1.3:2.7 molar ratio under ball milling for 18 h. Powder mixtures were sintered at 1600 °C for 4 h under Ar atmosphere. After grinding, 2 g of $Mo_2Ti_2AlC_3$ powder was immersed in 20 ml of HF (concentrated, 50%) under 200 rpm stirring at 55 °C for 90 h. Then the mixtures were washed many times by adding DI water and dry in vacuum oven. Ordered $Mo_2Ti_2C_3T_x$ was resoundingly prepared and base on DFT results predict that at 0 K, at least 26 ordered, solid solution MAX phases should be stable existing. In addition, base on simulating computation results, solid solution MXenes are also anticipated have high specific capacities for multiple valence elements such as Ca^{2+} [35, 36], Mg^{2+} and Al^{3+} [35]. Hence, polyphase solid solution Mxene has incomparable advantages over the ordinary ternary phase MXene in SC applications. This work provides a new route to add high-capacity elements to MXene to synthesize new energy storage materials with high electrochemical performance.

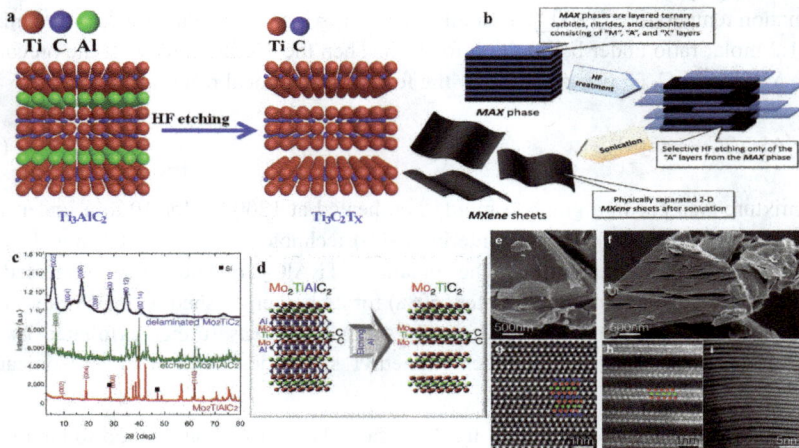

Figure 1.   Schematic of the exfoliation process for Ti_3AlC_2. a) Ti_3AlC_2 structure. [reprinted with permission from ref. 32. Copyright The Royal Society of Chemistry 2017.] b) Al atoms replaced by OH after reaction with HF. [reprinted with permission from ref. 10. Copyright 2012, American Chemical Society.] c) Breakage of the hydrogen bonds and separation of nanosheets after sonication in methanol. Synthesis and structure of Mo_2TiC_2 and $Mo_2Ti_2C_3$. (a) XRD patterns of Mo_2TiAlC_2 before (red) and after (green) HF treatment and after delamination (blue). In the delaminated sample, only c-direction peaks, (00l) peaks, are visible, corresponding to a c lattice parameter of 30.44 A°; the (110) peak is no longer observed, showing loss of order in nonbasal directions. (c) Schematics of Mo_2TiAlC_2 to $Mo_2TiC_2T_x$ transformations; red, green, blue, and black circles represent Mo, Ti, Al, and C atoms, respectively. (e and f) SEM images of Mo_2TiAlC_2 and $Mo_2TiC_2T_x$, respectively. Note the layers become open after etching in $Mo_2TiC_2T_x$. (g and h) HRSTEM of Mo_2TiAlC_2 and $Mo_2TiC_2T_x$, respectively. Atoms are shown with the same colors as (c). Atomic ordering is confirmed by EDX mapping. No Al was observed in EDX of $Mo_2TiC_2T_x$. (i) Lower magnification TEM image of (f) showing the layered structure throughout the sample. [reprinted with permission from ref. 34. Copyright 2015 American Chemical Society].

3.2   Element doped MXenes

The specific capacitance of MXenes can be further improved by bring in heteroatoms, such as nitrogen, sulfur, phosphorus and so on. The theories that illustrate the advantages of N-doped on graphene's electrochemical properties contain the Faraday capacitance behaviour of multiple functional groups at edges of the graphene basal plane [37-39], and the quaternary N (N-Q) groups can improve the conductivity of graphene [40, 41]. For the same reason, introducing N atoms into TiO_2 has been compare with undoped TiO_2, N-doped TiO_2 shown smaller band gap i.e. better conductivity [42-44]. Hence, N-doped of $Ti_3C_2T_x$ MXene is also expected to adjust the band gap in the MXenes, to improve the conductivity of Mxene and further enhance its electrochemical performance.

Wang and co-authors [45] reported a facile method to synthesis nitrogen doped $Ti_3C_2T_x$ MXene. In this work, annealing $Ti_3C_2T_x$ MXene in ammonia under 200 to 700 °C to obtain nitrogen doped MXenes with 1.7–20.7 at% surface N concentrations. N was introduced into MXene to replace C atoms, which greatly increased the layer spacing between $Ti_3C_2T_x$ Mxene plates. The nitrogen doped $Ti_3C_2T_x$ MXene demonstrated excellent electrochemical performance. After nitrogen doping, the specific capacitance of $Ti_3C_2T_x$ MXene increased by 4.6 times. Bao *et al.* [46] demonstrated a highly crumpled N-doping $Ti_3C_2T_x$ MXene as the electrode materials application on SCs. Crumpled N-doping $Ti_3C_2T_x$ MXene were obtained by heating melamine, which is the nitrogen source, can be electrostatic inserted into $Ti_3C_2T_x$ MXene nanosheets. The N-doping $Ti_3C_2T_x$ MXene exhibiting hierarchical pore structure with large specific surface area of 385 m^2 g^{-1} and a pore volume of 0.342 cm^3 g^{-1}, which are much higher than reported $Ti_3C_2T_x$ MXene, indicating excellent electrochemical performance.

3.3   MXene-based nanocomposites

The MXenes are widely used as electrode materials application on SCs with excellent cycling performance. However, the specific capacitance of MXene electrode materials is still very low, which is still far from practical application. In order to make MXene electrode material meet the practical application, its specific capacitance needs to be further improved. An effective way is to synthesize MXene-based composite materials to exert the synergistic effect of the two components.

Zhao *et al.* [47] reported a facile approach to assemble a flexible, sandwich-like structure MXene/CNT composite paper composite by a vacuum filtration route. The onion-like carbon/MXene composite and reduced graphene oxide/MXene composite were also synthesized and compared with MXene/CNT composite. The sandwich-like hierarchical structure electrodes demonstrated significantly enhanced electrochemical performance compared to those of pristine MXene and randomly mixed MXene/carbon films. This work provides a good method to assemble electrode materials for wearable supercapacitor devices. Zheng *et al.* [48] reported a MXene/polymer composite as electrode materials for SC application (Figure 2). $Ti_3C_2T_x$ MXene was mixed with either a charged polydiallyldimethylammonium chloride (PDDA) or an electrically neutral polyvinyl alcohol (PVA) to obtain $Ti_3C_2T_x$/polymer nanocomposites. The as-prepared $Ti_3C_2T_x$/PVA composites are flexible and exhibit high conductivity high to 2.2×10^4 S m^{-1}, much higher than the pure $Ti_3C_2T_x$ films, which shows 2.4×10^5 S m^{-1} conductivity. This MXene/polymer composite material is an ideal electrode material for supercapacitors attribute to its admirable electrical conductivity and excellent chemical stability behaviour. Mayerberger *et al.* [49] investigated MXene/polymer nanofibers composite fabricated by electrospinning approach. In this work, poly acrylic acid (PAA), polyethylene oxide (PEO), poly vinyl alcohol (PVA), and alginate/PEO were electrospun with $Ti_3C_2T_x$ MXene nanosheets. This work provides a new method to synthesizing MXene-based composites for microflexble SC devices.

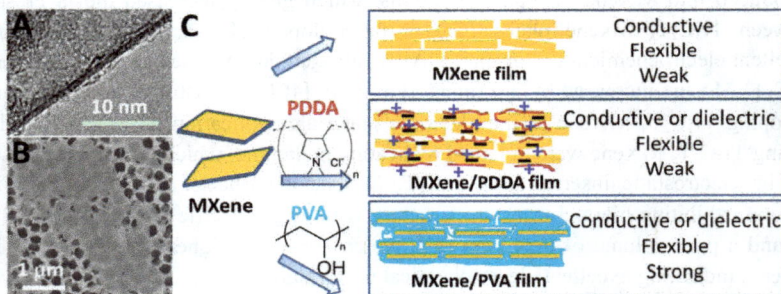

Figure 2.   (A) TEM and (B) SEM images of MXene flakes after delamination and before film manufacturing. (C) A schematic illustration of MXene-based functional films with adjustable properties. [reprinted with permission from ref. 48. Copyright 2014 National Academy of Sciences]

Zou *et al.* [50] prepared a MXene/Ag composites by direct degradation of a $AgNO_3$ aqueous solution in MXene. In this study, 0.1 g MXene was distracted in 80 mL of DI water and under sonication for 0.5 h and then stirred for 10 min to prepare a suspension. Afterwards, 0.05 g of $AgNO_3$ was dissolved in 20 mL of DI water as the Ag source and then slowly injected the $AgNO_3$ solution into the MXene suspension under 10 min reaction. The obtained MXene/Ag suspension was washed many time and dried in vacuum at 80 °C for 24 h. The average sphere size of Ag is about 35.2 nm, uniformly distributed in among of MXene layers.

Materials Research Forum LLC
doi: https://doi.org/10.21741/9781644900253-6

Zhu *et al.* [51] showed a TiO_2 nanoparticles anchored on $Ti_3C_2T_x$ MXene surface by a facile in situ hydrolysis and heat-treatment route. In a typical preparation of TiO_2/Ti_3C_2 nanocomposite, 0.15 g of $Ti_3C_2T_x$ MXene was dispersed in 200 mL ethanol under stirring to prepare a uniform suspension. The suspension was mixed with 0.5 mL of 0.4 mM KCl solution under stirring for 20 min and then injected 1 mL tetrabutyltitanate (TBOT) into mixture and stirred for 6 h. Finally, the TiO_2/MXene composite was obtained by washing and drying process.

Xia *et al.* [32] introduced a NiO/MXene nanocomposite by a hydrothermal method. In this study, 1.45 g $Ni(NO_3)_2 \cdot 6H_2O$ (0.1 mol), 0.18 g hexamethylenetetramine (0.025 mol), and 0.1 g $Ti_3C_2T_x$ MXene powder were dispersed in 50 mL DI water under stirring for 30 min to obtain uniform suspension, then injected into a 50 mL Teflon autoclave, and heated at 90°C for 4 h. The $Ni(OH)_2/Ti_3C_2T_x$ composite was washed with DI water and ethanol and then annealing at 400°C for 1 h to prepare $TiO_2/C-Ti_3C_2T_x$-MXene/NiO nanocomposite. In this work, NiO nanosheets uniformly cover on MXene surface.

Atomic layer deposition (ALD) is an advanced thin film deposition technology. The advantage of ALD is that it can precisely control the thickness of thin film and is widely used in the microelectronics industry. In 2017, Ahmeda *et al.* [52] synthesized SnO_2/MXene nanocomposites by ALD technology. The ALD deposition of SnO_2 and HfO_2 films were carried out on MXene surface, which kept on a Copper foil. The MXene film was prepared by mixing the MXene powder, acetylene black, and poly as a weight ratio of 80:10:10 in NMP solvent. The obtained slurry was equably coated on a Cu foil with 100 μm thicknesses by a doctor blade and dried at 80 °C for 24 h in vacuum oven. The MXene/Cu foil film was placed in ALD chamber for SnO_2 and HfO_2 coating. SnO_2 deposition was performed at two temperatures (150 °C or 200 °C). In the same way, HfO_2 films were deposited on the as-obtained MXene/SnO_2 composite films at 180 °C. It was successfully synthesized MXene/SnO_2 composite and MXene/SnO_2/ HfO_2 composite by this ALD method.

3.4   MXene quantum dots

As new members of MXene group, MXene quantum dots (MQDs) have attracted immensely attentions due to their special properties and widely used in the fields of optoelectronics, biotechnology, and energy. Theoretical and experimental investigations have shown that forming quantum dots (QDs) is an effective way to exploit novel physical properties due to quantum confinement, size and surface effects when they are made atomically thin [53, 54]. Compared with nanoflakes, QDs have ultra-small size and high surface activity, making them more suitable for energy storages.

Xue *et al.* [55] developed a water-soluble, monolayered Ti_3C_2 MQDs through a facile hydrothermal method. Ti_3AlC_2 (0.5 g) powder was slowly poured into 10 mL hydrofluoric acid (concentrated, 48%) and stirred at 60 °C for 20 h for MXene exfoliation. Then the suspension was washed many times to remove the HF residue, and dried at 80 °C for 12 h in a vacuum oven. The obtained Ti_3C_2 MXene was then placed in 20 mL DI water followed sonication for a while under an N_2-protected atmosphere. The pH of mixer was adjusted around 9 by ammonia and the mixer was injected in 75 mL autoclave at 100 °C, 120 °C and 150 °C for 6 h. The MXene Quantum Dots could be obtained by filtering the mixture by 220 nm membrane followed by concentrating using rotary evaporator under reduced atmosphere pressure.

In additon, Yu *et al.* [56] presented an effective fluorine-free method to synthesis Titanium carbide MXene quantum dots. The preparation of MXene QDs was based on a facile liquid exfoliation technique involving ultrasound probe sonication followed by bath sonication of powders of bulk Ti_3AlC_2. In brief, Ti_3AlC_2 powders (25 mg), Tetrabutylammonium Hydroxide (TBAOH, 10 mL), and water (25 mL) were added into a sealed conical tube (with a volume of 50 mL). The mixture was sonicated with a sonic tip for 6 h at a power of 480 W. The dispersions were further bath sonicated continuously for another 10 h at a power of 300 W. The suspension was centrifuged for 20 min at 7,000 rpm, and then carefully collected the supernatant containing MXene QDs.

In this section, the synthesis methods of MXene, element doped MXene, MXene-based nanocomposites, and MXene quantum dots are introduced. Among of those MXenes, N-doped MXenes, and MXene-based nanocomposites were used as electrode materials in SC. Although there is no report about MXene quantum dots in SC applications, it has great potential as electrode materials for energy storage. The new MXene-based electrode materials are also expected to be synthesized by simpler and safer methods.

4.   Structures, properties and supercapacitor applications

4.1   Single/few-layered MXene-based supercapacitors

The various as-prepared MXenes are endowed with the rare combination for 2D materials of excellent electronic conductivity comparable to multi-layered graphene and high hydrophilicity with small contact angles measured [10]. Understanding both the surface terminations and the interlayer interactions is the prerequisite to purposefully manipulate the properties of MXenes for their various applications. Consequently, modeling and simulation, such as DFT and to a lesser extent, molecular dynamics (MD) simulations, have been extensively applied to study the structure, to predict properties and to analyze potential applications of MXenes, guiding much of the experimental research in the field of MXenes, although most of the earlier modelling studies are regarding uniform terminating species [57-63].

Based on the DFT calculations, the 2D MXene flakes are quite firm when pulled along the basal plane. In addition, MXenes also have the same excellent conductivity as graphene. Compared with grapheme or black phosphorus, MXenes offers significant potential for composition regulation. The controllability of this component increases the potential applications range of MXene, making it possible to artificially regulate the properties of MXene [64]. Profit from their 2D nature, highly defined geometry, good conductivity, and high surface areas, MXenes have been exhibited as prospective electrode materials for SCs. The possibility of using MXenes as electrode materials for SC was first investigated by Lukatskaya *et al.*, [65] who reported titanium carbide MXene in various electrolytes. In this study, Na^+, K^+, NH^{4+}, Mg^{2+}, and Al^{3+} et al. cations also have been explored as intercalation ions, offering high capacitance higher than 300 F cm^{-3} (much higher than that of carbon materials, which less than 60 F cm^{-3}). This study provides a basis for study a large group of 2D MXenes in SC applications employing various electrolyte ions. To further understand how electrolyte ions affect the electrochemical properties, MXene paper was prepared by vacuum filtration route. The flexible, hydrophilic, additive-free, and conductive MXene with a specific surface area of 98 m^2 g^{-1}, and performed in KOH electrolyte. The CV curves of MXene paper shown a rectangular shape, indicating a highly reversible behaviors. Moreover, the EIS results demonstrated that the $Ti_3C_2T_x$ paper-based SCs were high conductivity than those multilayer $Ti_3C_2T_x$ MXene-based paper electrodes. The influencing factors of electrochemical properties can be summarized as follows: 1) the existence of a binder in the system; 2) loose contact between the restacked sheets in the MXene paper; 3) inaccessibility of the structure; and 4) the electrodes are too thick.

Moreover, Hu *et al.* [66] reported a binder-free $Ti_3C_2T_x$ MXene film in 1 M H_2SO_4 electrolyte that demonstrated a specific volumetric capacitance of 226 F cm^{-3} (2 mV s^{-1}) [66]. However, most of the electrolyte solutions uses in literature studies are consisting of dangerous toxic organic and inorganic chemicals like KOH and H_2SO_4. If leakage of the electrolyte occurs during the use it causes significant harm to human skin. Moreover, most of the electrode materials are unstable in strong acids or alkaline environments. Furthermore, organic electrolytes are currently not eco-friendly and their prices are relatively high. However, aqueous electrolytes based on lithium, sodium, potassium, and magnesium salts provide higher ionic conductivity, faster ion transport kinetics, and thereby can impart safety and economic benefits to SCs [67]. In general, Na^+ ions are abundantly available on earth; therefore, they can be easily obtained from seawater, which, in fact, contains 0.46 M of NaCl [68]. The SC fabrication process in the 0.5 M

NaCl as the electrolyte solution can be easier and cheaper owing to its abundancy, electrochemically stable, and eco-friendly character. Hence, Xia *et al.* [69] reported Seawater as electrolyte with high volumetric capacitance of MXene-based symmetric supercapacitors (SSC.

The field-emission SEM image of the obtained $Ti_3C_2T_x$ MXene is illustrated in Figure 3a, where the surface of $Ti_3C_2T_x$ MXene, after HF treatment, is typically exfoliated into several parallel flakes so called MXene-signature. The 2D structure of $Ti_3C_2T_x$ MXene observed using TEM (Figure 3b) image confirms a single-layered $Ti_3C_2T_x$ MXene with a measured thickness of 0.93 nm, which is consistent with the reported value [70]. The corresponding SAED pattern, shown in the inset of Figure 3b, supports for the 2D hexagonal crystalline structure of Mxene. Figure 3c shows the XRD patterns of Ti_3AlC_2 MAX phase before and after HF etching. After HF treatment, the Al layer has selectively been removed from Ti_3AlC_2, and the (002) and (004) peaks of Ti_3AlC_2 are shifted toward lower angles, demonstrating larger d spacing in $Ti_3C_2T_x$ than in Ti_3AlC_2 [10]. Brunauer-Emmett-Teller (BET) testing were performed to confirm the porosity of $Ti_3C_2T_x$ MXene (Figure 3d) where a typical type-IV isothermal with an H3-type hysteresis loop behaviour are noted. Both demonstrate the presence of slit-like pores in the $Ti_3C_2T_x$ MXene. The specific surface area, pore size, and total pore volume values of $Ti_3C_2T_x$ MXene are 10.0 m^2 g^{-1}, 24.1 nm, and 0.02 cm^3 g^{-1}, respectively. These slit-like pores of the $Ti_3C_2T_x$ MXene act as pore channels to promote rapid electrolyte transportation in the $Ti_3C_2T_x$ MXene electrode materials, which otherwise can enhance an electrochemical performance. XPS spectrum was utilized to characterize the valence states of the elements in $Ti_3C_2T_x$ MXene (Figure 4) where survey spectrum of $Ti_3C_2T_x$ MXene confirms the existence of Ti, C, O, and F elements. Presence of the O and F signals indicates an involvement of O, –OH and –F functional groups [70]. In Ti $2p$ spectrum (Figure 4b), Ti 2p core level was consists of two doublets (Ti $2p_{3/2}$-Ti $2p_{1/2}$) with a fixed area ratio equal to 2:1 and a doublet separation of 5.7 eV. And the Ti $2p_{3/2}$ components located at 455.1 eV and 456.3 eV were assigned to Ti-C and Ti-X, respectively [71, 72]. The Ti $2p_{1/2}$ components located at 460.9 eV and 462.8 eV were due to C-Ti-O_x and Ti-O, respectively [73]. The O 1S spectrum (Figure 4c) was fitted using 2 components peaks at 529.9 eV and 531.9 eV corresponding to O 1s in TiO_2 and hydroxyl groups [74]. The Ti-C and C-C bonds were identified in the C 1s (Figure 4d) spectrum at 281.3 eV and 284.6 eV [71, 75]. Therefore, based on the charge storage mechanism of $Ti_3C_2T_x$ MXene, multiple valence states and abundant functional groups of $Ti_3C_2T_x$ MXene can enable the electrolyte to wet electrode more easily by forming more ionic defects and can provide a short path for ion diffusion for better electrochemical performance [76].

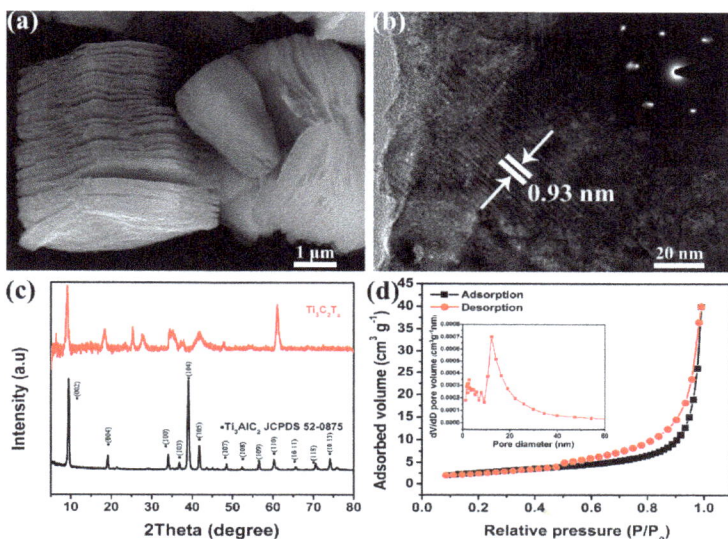

Figure 3. *(a) SEM image of $Ti_3C_2T_x$; (b) TEM image and corresponding SAED pattern (inset); (c) XRD patterns of Ti_3AlC_2 and $Ti_3C_2T_x$; (d) N_2 adsorption-desorption isotherms and pore-diameter distribution (inset) of $Ti_3C_2T_x$. [reprinted with permission from ref. 69. Copyright The Royal Society of Chemistry 2018.]*

CV curve is normally used to characterise the capacitive behaviour of an electrode material. Figure 5a shows the CV curves of $Ti_3C_2T_x$ MXene at various scan rates in NaCl electrolyte. All the CV curves retain nearly rectangular loops, which are caused by the electro-sorption/desorption of NaCl electrolyte cations (Na^+). Specific capacitance values of $Ti_3C_2T_x$ MXene electrode calculated from the GCD (GCD) curves (Figure 5b) are 67.7, 66.0, 46.0, 30.7, 20.0, and 10.0 F g^{-1} at the current densities of 1, 2, 3, 4, 5, and 10 A g^{-1}, respectively (Figure 5c). Specific capacitance reaching 67.7 F g^{-1} at the current density of 1 A g^{-1}, corresponding to the volumetric specific capacitance of 121.8 F cm^{-3}. Since high life-cycle stability is one of the important requirements while designing of SCs, $Ti_3C_2T_x$ electrode was subjected for 5000 charge/discharge cycles at a high current density of 10 A g^{-1}. The $Ti_3C_2T_x$ electrode still shows nearly 96.6% retention in its specific capacitrance (Figure 5d) which is supporting to an excellent structural stability of the $Ti_3C_2T_x$ electrode in 0.5 M NaCl electrolyte. The CV curves of the SSC at various scan rates in the potential range of 0 to 0.6 V in Figure 6 [77-81] exhibit a typical

MXenes: Fundamentals and Applications   Materials Research Forum LLC
Materials Research Foundations **51** (2019) 137-174   doi: https://doi.org/10.21741/9781644900253-6

rectangular profile indicating the high reversibility of the $Ti_3C_2T_x$ electrode material. The volumetric capacitance of SSC calculated from the GCD curves (Figure 6b) is presented in Figure 6c. The SSC device demonstrates as high as 27.4 F cm^{-3} volumetric capacitance at a low current density of 0.25 A g^{-1}, and 12.6 F cm^{-3} volumetric capacitance at a high current density of 3 A g^{-1}. The cycling stability of the SSC tested at a current density of 3 A g^{-1} (Figure 6d) confirms an excellent cycling performance with a capacitance loss of only 6.7% even after 5000 cycles of scans. In order to evaluate the performance of the $Ti_3C_2T_x$ MXene electrode in SC applications, SSC device was assembled by using $Ti_3C_2T_x$ MXene as the electrode material and is shown, as schematically illustrated, in Figure 6e. The energy density and power density values are important to assess the industrial potential of assembled SSC devices. A comparison of Ragone plots of the $Ti_3C_2T_x$ SSC device with literature data is shown in Figure 6f. The $Ti_3C_2T_x$ SSC device with a cell voltage of 0.6 V, in simulating seawater electrolyte, deliver volumetric energy densities of 1.74×10^{-3} Wh cm^{-3} at a power density of 0.15 W cm^{-3} and 0.68×10^{-3} Wh cm^{-3} at a power density of 1.53 W cm^{-3}.

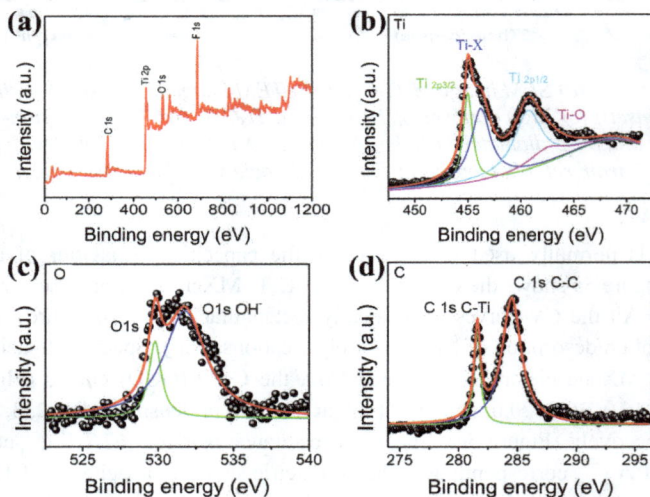

Figure 4.   (a) XPS survey spectrum. (b) Ti 2p, (c) O 1s, and (d) C 1s spectrums of
Ti$_3$C$_2$T$_x$ MXene. [reprinted with permission from ref. 69. Copyright The Royal Society of
Chemistry 2018.]

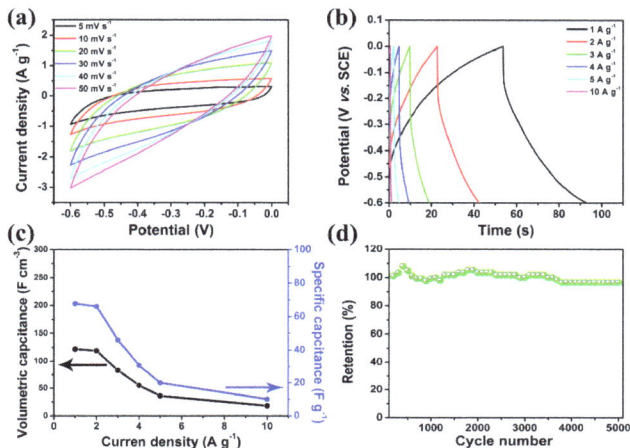

Figure 5.   *(a) CV curves at different scan rates; (b) GCD curves at various current densities; (c) the specific gravimetric capacitances (blue line) and volumetric capacitance (black line) at different current densities; (d) Cycling testing at a current density of 10 A g^{-1} of the Ti$_3$C$_2$T$_x$ electrode. [reprinted with permission from ref. 69. Copyright The Royal Society of Chemistry 2018.]*

Ti$_3$C$_2$T$_x$ MXene has demonstrated a high specific volumetric capacitance in aqueous electrolytes, but the potential window is very low (less than 1.23 V), which limits its performance. Lin *et al.* [82] tested Ti$_3$C$_2$T$_x$ MXene materials in organic electrolytes with a wide potential window. The obtained Ti$_3$C$_2$T$_x$ MXene hydrogel film prepared by vacuum filtration for use as SC electrodes tested in organic liquid electrolyte with a specific capacitance of 70 F g^{-1} (at 20 mV s^{-1} scan rate) in a 3 V potential window. This work explored the possibilities of using organic electrolytes for MXene-based SCs with high energy density.

In 2012, on the basis of DFT simulation results, Tang and co-workers [83] demonstrated that the F ion in MXene has a negative effect on the electrochemical performance of Li-ion batteries. Hence, R. B. Rakhi *et al.* [84] explored the influence of annealing on the electrochemical properties of MXene base electrode materials. In this work, influence of the annealing atmosphere (argon, nitrogen, N$_2$/H$_2$ forming gas, and air) on the structure and electrochemical performance of the MXene flakes was studied in detail. Compared with un-annealed treatment MXene, the annealed in air MXene flakes shown significant change on structure, morphology, and electrochemical properties. On the other hand, the MXene samples under argon, nitrogen, N$_2$/H$_2$ forming gas environment annealing process still maintain the original appearance. However, the obvious enhancement in specific capacitance was observed after argon, nitrogen, N_2/H_2 forming atmosphere annealing treatment. The highest specific capacitance and highest rate performance were obtained under N_2/H_2 atmosphere heating treatment. The enhanced electrochemical performance is attributed to highest carbon content, and lowest fluorine content on the surface of the sample after annealing treatments, and retaining the original 2D layered morphology and offering approachable ion diffusion path for the aqueous electrolyte.

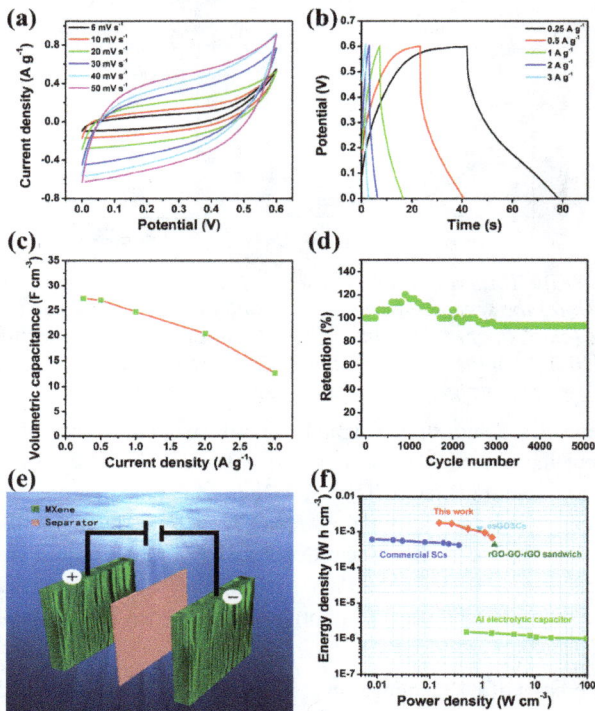

Figure 6.   (a) CV curves at various scan rates; (b) GCD curves; (c) Specific capacitance at different current densities; (d) Cycling testing at a current density of 3 A g^{-1}; (e) Schematic of device configuration of $Ti_3C_2T_x//Ti_3C_2T_x$ SSC. (f) Ragone plots showing comparative analysis of present work i.e. $Ti_3C_2T_x//Ti_3C_2T_x$ SSC with the Al-electrolytic capacitor [77], rGO-GO-rGO sandwich-based SCs [78], esGOSCs-based SCs [79], and commercial SCs [80]. [reprinted with permission from ref. 69. Copyright The Royal Society of Chemistry 2018.]

Volume performance makes sense for portable electronics and electric vehicles. Ghidiu *et al.* [85] developed a method of preparing MXene by a solution of LiF and HCl exfoliation process (Figure 7a). The resulting hydrophilic material with clay-like shape was rolled into films of tens of micrometers thick. The electrochemical performances of additive-free films were tested in KOH electrolyte exhibited high specific volumetric capacitances high to 900 F cm^{-3} (Figure 7b), with excellent stability and rate performances (Figure 7c). The electrodes showed excellent cycling stability even after 10,000 cycles (Figure 7d). In addition, coulombic efficiency is close to 100% (inset in Figure 7d), indicating that the excellent electrochemical performances is not attributable to parasitic reactions. This work provides a fast method for the synthesis of MXene films, while avoiding the use of highly toxic HF as the etching agent.

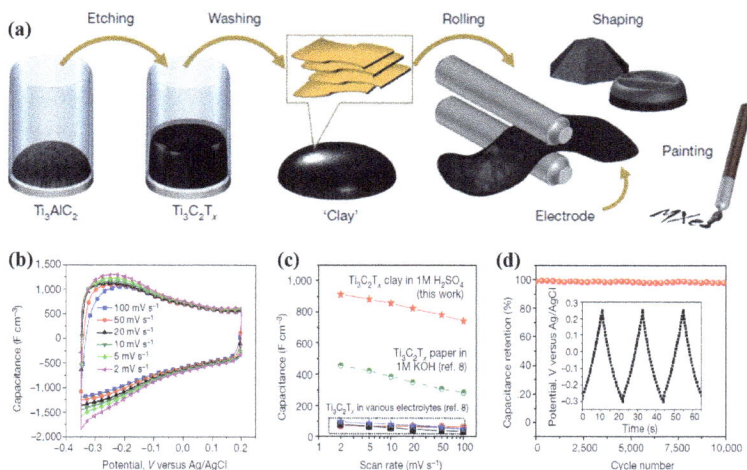

Figure 7.   Schematic of MXene clay synthesis and electrode preparation. (a) MAX phase is etched in a solution of acid and fluoride salt (step 1), then washed with water to remove reaction products and raise the pH towards neutral (step 2). The resulting sediment behaves like a clay; it can be rolled to produce flexible, freestanding films (step 3), moulded and dried to yield conducting objects of desired shape (step 4), or diluted and painted onto a substrate to yield a conductive coating (step 5). (b) Cyclic voltammetry profiles at different scan rates for a electrode (thickness: 5 mm) in 1M H$_2$SO$_4$. (c) Comparison of rate performances reported in this work and previously for HF-produced MXene [65]. (d) Capacitance retention test of a rolled electrode (thickness: 5 mm) in 1M H$_2$SO$_4$. Inset shows galvanostatic cycling data collected at 10 A g^{-1} [81]. [reprinted with permission from ref. 85. Copyright 2018, Nature.]

Most of the previous studies focused on $Ti_3C_2T_x$ MXenes. So far, more than 20 have been successfully produced. In these MXenes, V_2CT_x is a promising electrode material due to its lightness and multiple valences. Previous work has shown that V_2CT_x MXene has the ability ions intercalation when used as an electrode material for lithium ion batteries [33, 86,87]. Shan *et al.* [88] introduced on the electrochemical properties of V_2CT_x MXene in various electrolytes. The transparent conductive V_2CT_x MXene film demonstrated high specific capacitances, specifically 487 F g^{-1} in 1 M H_2SO_4, 225 F g^{-1} in 1 M $MgSO_4$, and 184 F g^{-1} in 1 M KOH electrolyte, which are higher than previously reported specific capacitance for few MXene film electrodes. This work demonstrated the electrochemical properties of V_2CT_x MXene in different electrolytes, and confirmed that V_2CT_x MXene obtained very high capacitance in acidic electrolytes.

Halim *et al.* [31] synthesized large scale 2D Mo_2CT_x MXene by selective etching gallium layers from the Mo_2Ga_2C. In this work, the electrical properties of Mo_2CT_x MXene at different temperatures were explored. The conductivity of Mo_2CT_x MXene films decreases exponentially with the decrease of temperature from 300 to 10 K, indicating semiconductor behavior of Mo_2CT_x MXene. In addition, the magnetic properties of MXene were also studied, at the temperature of 10 K, the magnetoresistance of Mo_2CT_x MXene is positive. The specific volumetric capacitance of 2 μm thick Mo_2CT_x "paper" in 1 M H_2SO_4 achieved 700 F cm^{-3} at scan rate of 2 mV s^{-1}. The Mo_2CT_x MXene films electrodes exhibited outstanding cycling performance, with almost no degradation after 10,000 cycles.

4.2 Element doped MXenes

Same as graphene, N-doping of $Ti_3C_2T_x$ is also expected to tune the electronic structure in the $Ti_3C_2T_x$ MXenes, leads to a narrower band gap which means better conducitvity. Wen *et al.* [45] demonstrated a facile method to synthesis N-doping $Ti_3C_2T_x$ MXene electrodes by heating $Ti_3C_2T_x$ MXene in ammonia. In this work, after annealing, $Ti_3C_2T_x$ MXene achieved a high nitrogen doping ratio, with 1.7–20.7 at% surface nitrogen concentrations from 200 °C to 700 °C ammonia-assistant annealing process. The N-$Ti_3C_2T_x$ electrodes exhibited a up to 1.6 times higher in specific capacitance than the undoped $Ti_3C_2T_x$ MXene electrodes, the reason may attributed to the N substitute C atoms result to the increasing of interlayer spacing between the $Ti_3C_2T_x$ MXene, and the narrowing band gap of $Ti_3C_2T_x$ MXene. After nitrogen doping treatment, the N-doped $Ti_3C_2T_x$ MXene materials demonstrated enhanced specific capacitance of 192 F g^{-1} in 1 M H_2SO_4 and 82 F g^{-1} in 1 M $MgSO_4$ electrolyte, which are significantly superior than those of the un-doped $Ti_3C_2T_x$ MXene electrodes (34 F g^{-1} in 1 M H_2SO_4 and 52 F g^{-1} in 1 M $MgSO_4$ electrolyte).

Moreover, Yang *et al.* [89] reported an enhanced specific capacitance SC electrode material based on N-doping $Ti_3C_2T_x$ MXene prepared by an urea-assisted heating process. The 2D N-doping $Ti_3C_2T_x$ MXene flakes demonstrated a high specific capacitance of 266.5 F g^{-1} (at 5 mV s^{-1}) in 6 M KOH aqueous electrolyte, with an excellent cycling performance. This new method synthesizes N-doping MXene as a promising electrode material for high-performance SCs.

4.3   MXene composites-based supercapacitors

Recently, many researchers have reported the synthesis of single or few layers of MXene in various approaches, and showed excellent electrochemical properties. However, there are still challenges in the practical application of MXene-baded SCs due to its unsatisfactory specific capacitance. To further improve the specific capacitance of MXene-based SCs, surface decoration of MXene with high theoretical specific capacitance materials to obtain a MXene-based nanocomposites is an important strategy.

Zhao *et al.* [47] designed sandwich-structure MXene/CNT composite flexible paper electrodes by a vacuum filtration route. The sandwich-like MXene/CNT composite papers used as electrodes of SC with outstanding higher specific volumetric capacitances and good rate performances compared with pure MXene and randomly mixed MXene/CNT papers. The sandwich-like MXene/SWCNT composite electrode exhibited a high specific volumetric capacitance of 390 F cm^{-3}, 350 F cm^{-3} at a scan rate of 2 mV s^{-1}, 5 A g^{-1}, respectively. The cycling performance of sandwich-like MXene/SWCNT demonstrated no degradation after 10000 cycles. Moreover, the sandwich-like MXene/rGO composite papers showed a high volumetric capacitance of 435 F cm^{-3} and excellent cycling performance. The sandwich-like electrodes demonstrated markedly enhanced electrochemical performance superior than those of pure MXene and unorderly mixed MXene/CNT composite-based electrodes.

Nickel oxide nanosheets are a promising electrode material with high theoretical capacitance. Xia *et al.* [32] introduced a $Ti_3C_2T_x$ MXene electrode decorated with nickel oxide nanosheets were prepared by a hydrothermal route. The nickel oxide nanosheets were grown and immobilized on the surface of $Ti_3C_2T_x$ MXene. The carbon-supported TiO_2 layer was derived from $Ti_3C_2T_x$-MXene during a thermal annealing process. This results in the synthesis of multiphase Mxene-based composites with NiO, C, TiO_2, and $Ti_3C_2T_x$ MXene, which named as NiO-grown derived-TiO_2/C-$Ti_3C_2T_x$-MXene nanocomposite (Ni-dMXNC). An electrode Ni-dMXNC showed a significantly specific capacity of 414.1 F cm^{-3}, 242.6 F cm^{-3} at current density of 1 A g^{-1}, 10 A g^{-1}, respectively. Moreover, an asymmetric supercapacitor (ASC) device utilized Ni-dMXNC, $Ti_3C_2T_x$ MXene as the positive electrode negative electrode, respectively, was exhibited an energy density of 1.04×10^{-2} Wh cm^{-3} and the corresponding power density of 0.22 W cm^{-3}, and cycling performance with 27.9 % degradation after 5000 cycles. The improved capacitance is due to the hierarchical 3D porous Ni-dMXNC structure and provides a shorten electrolyte ion diffusion path.

Beside nickel oxide, manganese dioxide is also a candidate material for supercapacitor electrode with high specific capacitance. Tian *et al.* [90] developed a flexible and free-standing $Ti_3C_2T_x$ file decorated with MnO_x nanoparticles as a high volumetric capacitance electrode for SC. This MnO_x-$Ti_3C_2T_x$ composite film hybridized nanoparticles of Mn_2O_3 and MnO on MXene surface showed electrochemical performances as electrodes of SCs with a high volumetric capacitance of 602.0 F cm^{-3} at scan rate of 2 mV s^{-1}. In addition, symmetric SCs based on the MnOx-$Ti_3C_2T_x$ composite electrodes exhibited a good energy capacity (13.64 mWh cm^{-3}) and a good cycling performance of 10.2% capacitance degradation after 10000 cycles. Therefore, these highly flexible and free-standing MnO_x-$Ti_3C_2T_x$ composite films are promising electrodes materials for wearable SCs device.

Layered double hydroxide (LDH) has attracted much attention as a novel and promising electrode material for SCs. Wang *et al.* [91] introduced a Ni-Al-LDH platelets decorated MXene surface composite, which is beneficial to exposing the active sites of LDH for electrolyte, further reduce the volume expansion of LDH during the charge/discharge process. The best condition of MXene/LDH composite exhibits a high specific capacitance of 1061 F g^{-1} (at 1 A g^{-1} current density), with moderate cycling performance of 70% capacitance retention after 4000 cycles. However, there is still a lack of detailed structural design for the most MXene-based SCs electrode materials. Nonetheless, for this LDH/$Ti_3C_2T_x$ composite, the real heterostructure between $Ti_3C_2T_x$ and LDH with an inadequate contact in this composite; and the wall thickness of LDH is still very thick. Therefore, Zhao *et al.* [92] shown a particular hierarchical Ni-Co-Al-LDH/$Ti_3C_2T_x$ composite with molecular-level nanoflakes are prepared by a facile negatively charged $Ti_3C_2T_x$ and positively charged Ni-Co-Al-LDH nanosheets approach. The LDH/$Ti_3C_2T_x$ composite demonstrated an excellent specific capacitance of 748.2 F g^{-1} at current density of 1 A g^{-1}, shown a remarkable rate performance. Moreover, this LDH/$Ti_3C_2T_x$ composite as positive electrode and active carbon as the negative electrode were fabricated as an all-solid-state flexible ASC device demonstrated a 45.8 Wh kg^{-1} energy density.

There are many researches on positive electrode materials, but few on negative electrode materials. Among of these negative electrode materials, bismuth oxychloride has attracted attention attributed to its high volumetric density, good electrochemical performance, multivalent, remarkable oxide ion conductivity, and high theoretical capacitance [93,94]. Xia *et al.* [95] developed a facile approach to prepare the bismuth oxychloride nanosheets anchored $Ti_3C_2T_x$ MXene composite (TCBOC) with 3D flower-like structure by a chemical bath deposition (CBD) route [95] (Figure 8). The SEM images of the $Ti_3C_2T_x$ MXene are shown in Figure 9. Figure 9a, b shows that the $Ti_3C_2T_x$ MXene with thickness of 50–200 nm. Single layer of $Ti_3C_2T_x$ MXene sheets with thickness of 1 nm [96], indicating that the obtained MXene can be called multilayer MXene. The functional groups (*i.e.*, =O, –OH, and –F) on the $Ti_3C_2T_x$ Mxene surface contribute to BiOCl adsorption. The SEM images of the TCBOC composite as shown in Figure 9c and d, exhibited that the tetragonal PbFCl-type structure BiOCl flakes with lattice constants a = 0.3890 nm and c = 0.7890 nm. The TCBOC tested as negative electrode demonstrated a 396.5 F cm^{-3} at 1 A g^{-1} and 228.0 F cm^{-3} at 15 A g^{-1} outstanding specific volumetric capacitance. The configuration of TCBOC composite based SSC as shown in Figure 10a [95, 98-102]. Figure 10b shows CV curves of the device at various scan rates, a pair of evident redox peaks to appear in the curves due to the faradic pseudocapacitance contribution. The GCD curves of the device at different current densities were demonstrated in Figure 10c and shown a nonlinear curve attributed to the redox reaction from BiOCl and modified $Ti_3C_2T_x$ MXene contributions. The specific volumetric capacitances (Figure 10d) of the TCBOC SSC device were 64.3 F cm^{-3} and 63.3 F cm^{-3} at current densities of 0.5 A g^{-1} and 1.0 A g^{-1}. Moreover, the TCBOC based SSC configuration device demonstrated excellent cycling stability with 15 degradation after 5000 cycles (at current density of 5 A g^{-1}) (Figure 10e). Figure 10f shows that the TCBOC SSC device has an maximal energy density of 15.2 Wh kg^{-1} and at 567.4 W kg^{-1} corresponding power density, and this device can achieved to maximal power density of 3,756.8 W kg^{-1} (at the energy density 6.2 Wh kg^{-1}). The improved electrochemical performance was due to the synergistic effect of BiOCl and $Ti_3C_2T_x$ MXene flakes in the TCBOC composite, which can be attributed as fllows: 1) Enhanced surface area when the BiOCl flakes grown on $Ti_3C_2T_x$ MXene surface; 2) The activities of BiOCl flakes and $Ti_3C_2T_x$ MXene provided abundant active sites; and 3) The excellent conductivity of a $Ti_3C_2T_x$ MXene material leading to faster transfer of electrons.

Zhao *et al.* [97] prepared N-doped carbon decorated $Ti_3C_2T_x$ MXene composite as negative electrode for supercapacitors. The N-doping carbon anchored on $Ti_3C_2T_x$ MXene surface, which named as $Ti_3C_2T_x@NC$ nanocomposites were prepared through an in-situ self-polymerization of dopamine on the surface of $Ti_3C_2T_x$ followed by a carbonization approach. The $Ti_3C_2T_x@NC$ nanocomposites exhibited high surface area and good conductivity. Benefit by these novel designed $Ti_3C_2T_x@NC$ nanocomposites electrode materials demonstrated a high specific capacitance of 442.2 F g^{-1} at a current density of 1 A g^{-1}, which is 2.81 times higher than that of pristine $Ti_3C_2T_x$ MXene. Cyclic

Materials Research Forum LLC
doi: https://doi.org/10.21741/9781644900253-6

stability is an important parameter of SCs, this $Ti_3C_2T_x@NC$ nanocomposite exhibited an remarkable cycling performance with 91.9% capacitance retention after 5000 cycles and a excellent rate capability of 92.5% at a high current density of 10 A g^{-1}. Moreover, the $Ti_3C_2T_x@NC$-based SSC showed an outstanding energy density and power density. Hence, the well synthesized $Ti_3C_2T_x@NC$ nanocomposites provided new exploration in the field of energy storage.

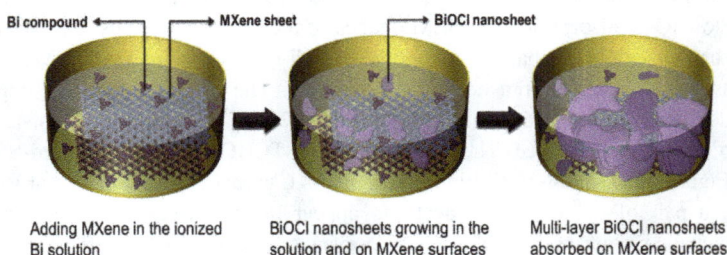

Figure 8. *The synthesis process of TCBOC via CBD method. [reprinted with permission from ref. 95. Copyright 2018, Elsevier.]*

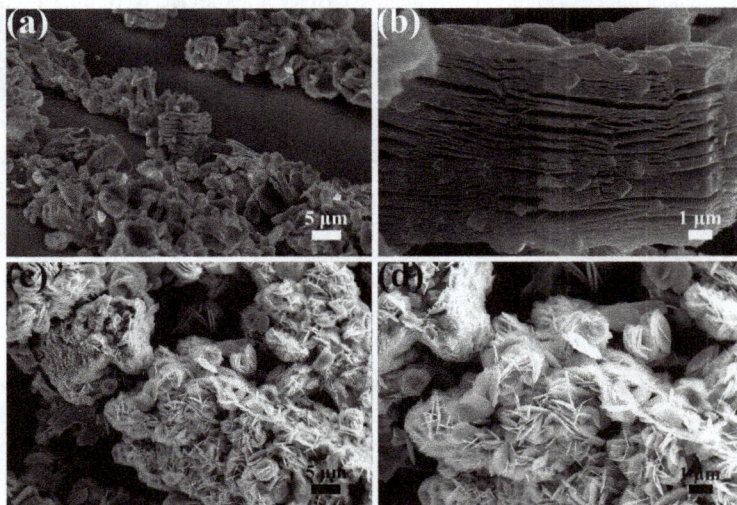

Figure 9. *(a) Low- and (b) high-magnification SEM images of $Ti_3C_2T_x$ MXene; (c) low-and (d) high-magnification SEM images of the TCBOC composite. [reprinted with permission from ref. 95. Copyright 2018, Elsevier.]*

Materials Research Forum LLC
doi: https://doi.org/10.21741/9781644900253-6

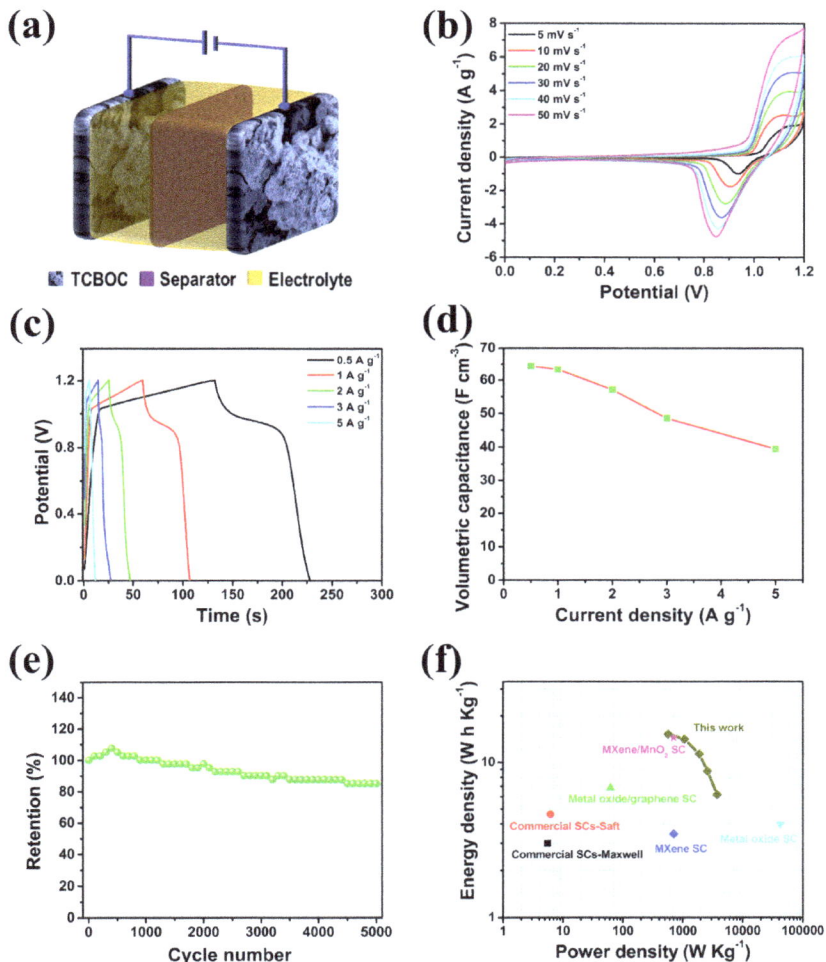

Figure 10.   (a) Schematic of the TCBOC SSC configuration. (b) CV curves of TCBOC SSC at different scan rates. (c) GCD curves of TCBOC SSC. (d) Specific capacitance of TCBOC SSC at different current densities. (e) Cycling performance of the TCBOC SSC at current density of 5 A g^{-1}. (f) Ragone plots of the TCBOC SSC compared with MXene [98], MXene/metal oxide composite [99], metal oxide SC [100], metal oxide/graphene SC [101], and commercial SC [102]. [reprinted with permission from ref. 90. Copyright 2018, Elsevier.]

Summary and outlook

In recent years, a lot of efforts have been put into developing electrode materials for high performance SCs. During the past decade, a lot of research on graphene has aroused great interest in the exploration of 2D materials. Therefore, in recent years, the discovery of similar 2D material MXene exfoliated from the lamellar MAX phase has attracted attention. Because of their novel properties, including low friction coefficient, good oxidation resistance, high electrical and thermal conductivity, hydrophilia, composition adjustable, high surface area, good thermal and chemical stability, and ability to host a broad range of intercalates, MXene materials have been employed as electrode materials for SCs and surpasses graphene in some electrochemical properties such as specific volumetric capacitance. In this work, recent progress in the development of 2D MXene materials for future SC applications were illustrated.

MAX phases are a very large group of ternary carbides and nitrides with more than 70 reported. Moreover, there are a lot of solid solutions and ordered double transition metal structures. DFT calculations results exhibited that for certain combinations of transition metals, ordered MXenes are more stable than their solid-solution duplicates, and exceed 25 ordered MXenes have been expected to exist [34]. Although more than 100 different components MAX phases have been predicted existence, there are many more layered MAX phases waiting to be converted to MXenes. Theoretical simulation results shown that many MXenes have not yet been successful exfoliated, such as Sc_2C, Hf_2C or W_2C [103]. Therefore, the exfoliation of new MAX phases and other laminar carbide and nitride precursors has become an important research direction. We expect there are still many MXenes that are more suitable as supercapacitor electrodes, and worth investigating.

Generally, most of MXene can be prepared by selectively removal of A from the MAX phase by hydrofluoric acid etching. Aside from hydrofluoric acid, other less toxic etchant mixtures, such as LiF+HCl, NaF+HCl, and NaOH et al. have also been successfully performed to selectively remove the A layer elements. The obtained MXene with abundant functional groups such as hydroxyl, oxygen, and fluorine lead to hydrophilic nature. However, more work is needed to prepare fluoride-free MXenes, which may provide new opportunities to further investigate the particular features of these MXene materials. Annealing is an effective method to reduce fluorine functional groups, the MXene heated in N_2/H_2 environmental demonstrated the highest specific capacitance of 51 F g^{-1} (at 1 A g^{-1}) and high rate performance (86%), due to the highest carbon content, which improved the conductivity of the MXene electrode, and lowest fluorine content on the surface of the MXene during the annealing in N_2/H_2 atmosphere.

Importantly, clay-like $Ti_3C_2T_x$ MXene employed a solution of LiF and HCl. The obtained hydrophilic material expands when hydrated and looks like clay, dried into highly conductive solids, or rolled into films with tens of microns thick. The additive-free clay-like MXene films showed ultra-high volumetric capacitance of 900 F cm^{-3}, with remarkable cycling stability and rate performances. Like graphene, nitrogen doping is an effective method to improve the electrochemical properties of MXenes. Methods of nitrogen doping include annealing and hydrothermal methods. Expect more elements to achieve MXene doping, such as sulfur and phosphorus.

With a large number of MXene-based composite are synthesized, with different filler(s), such as metal oxides with higher specific capacity, for energy storage applications, is inevitable, due to the inherent moderate theoretical capacity of MXenes. Incorporation of interlayer spacers, such as graphene, CNTs or PPy, to provide additional electron conductive pathways and maintain interlayer spaces that promote electrolyte ion access and transportation, has been successful improved the overall electrochemical performances of the materials. It remains intriguingly challenging to design a hierarchical structure with high conductivity and good chemical stability.

This novel large family of 2D materials exhibits huge potential as promising electrode materials for SC applications, encourage further explorations. Among of the MXene-based materials, MXene quantum dots attract attention due to their special properties based on its own unique quantum effects, when the particle size in nanometer level, size limit field will cause the surface effect and tunneling effect, size effect, and showed a lot of different from macro material physical and chemical properties, is expected to be in optics, electricity, magnetism, catalysis, biological medicine and functional material widely used. So far, there are few reports of MXene quantum dots application in the area of energy storage. However, in view of the continuous development of graphene and black phosphorus quantum dots and their applications in the field of energy storage, it is believed that the application of MXene quantum dots in the direction of energy storage is not far off.

References

[1] S. Hosogai, H. Tsutsumi, Electrospun nickel oxide/polymer fibrous electrodes for electrochemical capacitors and effect of heat treatment process on their performance, J. Power Sources 194 (2009) 1213-1217.
https://doi.org/10.1016/j.jpowsour.2009.06.044

[2] P. Simon, Y. Gogotsi, Materials for electrochemical capacitors, Nat. Mater. 7 (2008) 845-854.

[3] P.C. Gao, W.Y. Tsai, B. Daffos, P.L. Taberna, C.R. Perez, Y. Gogotsi, P. Simon, F. Favier, Graphene-like carbide derived carbon for high-power supercapacitors, Nano Energy 12 (2015) 197-206. https://doi.org/10.1016/j.nanoen.2014.12.017

[4] S.Y. Zhang, L. Ren, S.J. Peng, Zn_2SiO_4 urchin-like microspheres: controlled synthesis and application in lithium-ion batteries, Crystengcomm 16 (2014) 6195-6202. https://doi.org/10.1039/c4ce00479e

[5] K. Novoselov, A. Mishchenko, A. Carvalho, A.C. Neto, 2D materials and van der Waals heterostructures, Science 353 (2016) aac 9439. https://doi.org/10.1126/science.aac9439

[6] C. Liu, Z. Yu, D. Neff, A. Zhamu, B.Z. Jang, Graphene-based supercapacitor with an ultrahigh energy density, Nano Lett. 10 (2010) 4863-4868. https://doi.org/10.1021/nl102661q

[7] Y. Wang, Z.Q. Shi, Y. Huang, Y.F. Ma, C.Y. Wang, M.M. Chen, Y.S. Chen, Supercapacitor devices based on graphene materials, J. Phys. Chem. C 113 (2009) 13103-13107. https://doi.org/10.1021/jp902214f

[8] C.G. Liu, Z.N. Yu, D. Neff, A. Zhamu, B.Z. Jang, Graphene-based supercapacitor with an ultrahigh energy density, Nano Lett. 10 (2010) 4863-4868. https://doi.org/10.1021/nl102661q

[9] M. Ghidiu, M.R. Lukatskaya, M.-Q. Zhao, Y. Gogotsi, M.W. Barsoum, Conductive two-dimensional titanium carbide/clay/'with high volumetric capacitance, Nature 516 (2014) 78-81. https://doi.org/10.1038/nature13970

[10] M. Naguib, O. Mashtalir, J. Carle, V. Presser, J. Lu, L. Hultman, Y. Gogotsi, M.W. Barsoum, Two-dimensional transition metal carbides, ACS Nano 6 (2012) 1322-1331. https://doi.org/10.1021/nn204153h

[11] M.W. Barsoum, The $M_{N+1}AX_N$ phases: a new class of solids: thermodynamically stable nanolaminates, Prog. Solid State Chem. 28 (2000) 201-281. https://doi.org/10.1016/s0079-6786(00)00006-6

[12] X.F. Yu, J.B. Cheng, Z.B. Liu, Q.Z. Li, W.Z. Li, X. Yang, B. Xiao, Mg intercalation into Ti_2C building block, Chem. Phys. Lett. 629 (2015) 36-39. https://doi.org/10.1016/j.cplett.2015.04.015

[13] M. Kurtoglu, M. Naguib, Y. Gogotsi, M.W. Barsoum, First principles study of two-dimensional early transition metal carbides, MRS Commun. 2 (2012) 133-137. https://doi.org/10.1557/mrc.2012.25

[14] M. Ghaffari, Y. Zhou, H. Xu, M. Lin, T.Y. Kim, R.S. Ruoff, Q. Zhang, High-volumetric performance aligned nano-porous microwave exfoliated graphite oxide-based electrochemical capacitors, Adv. Mater. 25 (2013) 4879-4885. https://doi.org/10.1002/adma.201301243

[15] J. Yan, C.E. Ren, K. Maleski, C.B. Hatter, B. Anasori, P. Urbankowski, A. Sarycheva, Y. Gogotsi, Flexible MXene/graphene films for ultrafast supercapacitors with outstanding volumetric capacitance, Adv. Funct. Mater. 27 (2017) 1701264-1701274. https://doi.org/10.1002/adfm.201701264

[16] H. Li, Y. Hou, F. Wang, M.R. Lohe, X. Zhuang, L. Niu, X. Feng, Flexible all-solid-state supercapacitors with high volumetric capacitances boosted by solution processable MXene and electrochemically exfoliated graphene, Adv. Energy Mater. 7 (2017) 1601847-1601853. https://doi.org/10.1002/aenm.201601847

[17] Y.-Y. Peng, B. Akuzum, N. Kurra, M.-Q. Zhao, M. Alhabeb, B. Anasori, E.C. Kumbur, H.N. Alshareef, M.-D. Ger, Y. Gogotsi, All-MXene (2D titanium carbide) solid-state microsupercapacitors for on-chip energy storage, Energy Environ. Sci. 9 (2016) 2847-2854. https://doi.org/10.1039/c6ee01717g

[18] J.C. Lei, X. Zhang, Z. Zhou, Recent advances in MXene: Preparation, properties, and applications, Front Phys-Beijing 10 (2015) 276-286.

[19] Y.J. Zhang, L. Wang, N.N. Zhang, Z.J. Zhou, Adsorptive environmental applications of MXene nanomaterials: a review, RSC Adv. 8 (2018) 19895-19905. https://doi.org/10.1039/c8ra03077d

[20] M. Naguib, V.N. Mochalin, M.W. Barsoum, Y. Gogotsi, 25th Anniversary Article: MXenes: A new family of two-dimensional materials, Adv. Mater. 26 (2014) 992-1005. https://doi.org/10.1002/adma.201304138

[21] Q.K. Hu, D.D. Sun, Q.H. Wu, H.Y. Wang, L.B. Wang, B.Z. Liu, A.G. Zhou, J.L. He, MXene: A new family of promising hydrogen storage medium, J. Phys. Chem. A 117 (2013) 14253-14260. https://doi.org/10.1021/jp409585v

[22] M. Naguib, J. Come, B. Dyatkin, V. Presser, P.L. Taberna, P. Simon, M.W. Barsoum, Y. Gogotsi, MXene: a promising transition metal carbide anode for lithium-ion batteries, Electrochem. Commun. 16 (2012) 61-64. https://doi.org/10.1016/j.elecom.2012.01.002

[23] M. Khazaei, A. Ranjbar, M. Arai, T. Sasaki, S. Yunoki, Electronic properties and applications of MXenes: a theoretical review, J. Mater. Chem. C 5 (2017) 2488-2503. https://doi.org/10.1039/c7tc00140a

[24] M. Winter, R.J. Brodd, What are batteries, fuel cells, and supercapacitors? Chem. Rev. 105 (2005) 1021-1021. https://doi.org/10.1021/cr040110e

[25] M.S. Halper, J.C. Ellenbogen, Supercapacitors: A brief overview, MITRE Nanosystems Group, Virginia, (2006).

[26] J. Fernández, T. Morishita, M. Toyoda, M. Inagaki, F. Stoeckli, T.A. Centeno, Performance of mesoporous carbons derived from poly (vinyl alcohol) in electrochemical capacitors, J. Power Sources 175 (2008) 675-679. https://doi.org/10.1016/j.jpowsour.2007.09.042

[27] Z. Li, J. Han, L. Fan, M. Wang, S. Tao, R. Guo, The anion exchange strategy towards mesoporous α-Ni(OH)$_2$ nanowires with multinanocavities for high-performance supercapacitors, Chem. Commun. 51 (2015) 3053-3056. https://doi.org/10.1039/c4cc09225b

[28] L.Q. Mai, A. Minhas-Khan, X. Tian, K.M. Hercule, Y.-L. Zhao, X. Lin, X. Xu, Synergistic interaction between redox-active electrolyte and binder-free functionalized carbon for ultrahigh supercapacitor performance, Nat. Commun. 4 (2013) 2923-2930. https://doi.org/10.1038/ncomms3923

[29] A. Laheäär, P. Przygocki, Q. Abbas, F. Béguin, Appropriate methods for evaluating the efficiency and capacitive behavior of different types of supercapacitors, Electrochem. Commun. 60 (2015) 21-25. https://doi.org/10.1016/j.elecom.2015.07.022

[30] M. Naguib, M. Kurtoglu, V. Presser, J. Lu, J.J. Niu, M. Heon, L. Hultman, Y. Gogotsi, M.W. Barsoum, Two-dimensional nanocrystals produced by exfoliation of Ti$_3$AlC$_2$, Adv. Mater. 23 (2011) 4248-4253. https://doi.org/10.1002/adma.201102306

[31] J. Halim, S. Kota, M.R. Lukatskaya, M. Naguib, M.Q. Zhao, E.J. Moon, J. Pitock, J. Nanda, S.J. May, Y. Gogotsi, M.W. Barsoum, Synthesis and characterization of 2D molybdenum carbide (MXene), Adv. Funct. Mater. 26 (2016) 3118-3127. https://doi.org/10.1002/adfm.201505328

[32] Q.X. Xia, J.J. Fu, J.M. Yun, R.S. Mane, K.H. Kim, High volumetric energy density annealed-MXene-nickel oxide/MXene asymmetric supercapacitor, RSC Adv. 7 (2017) 11000-11011. https://doi.org/10.1039/c6ra27880a

[33] M. Naguib, J. Halim, J. Lu, K.M. Cook, L. Hultman, Y. Gogotsi, M.W. Barsoum, New two-dimensional niobium and vanadium carbides as promising materials for Li-ion batteries, J. Am. Chem. Soc. 135 (2013) 15966-15969. https://doi.org/10.1021/ja405735d

[34] B. Anasori, Y. Xie, M. Beidaghi, J. Lu, B.C. Hosler, L. Hultman, P.R.C. Kent, Y. Gogotsi, M.W. Barsoum, Two-dimensional, ordered, double transition metals carbides (MXenes), ACS Nano 9 (2015) 9507-9516. https://doi.org/10.1021/acsnano.5b03591

[35] Y. Xie, Y. Dall'Agnese, M. Naguib, Y. Gogotsi, M.W. Barsoum, H.L.L. Zhuang, P.R.C. Kent, Prediction and characterization of mxene nanosheet anodes for non-lithium-ion batteries, ACS Nano 8 (2014) 9606-9615. https://doi.org/10.1021/nn503921j

[36] D.Q. Er, J.W. Li, M. Naguib, Y. Gogotsi, V.B. Shenoy, Ti_3C_2 MXene as a high capacity electrode material for metal (Li, Na, K, Ca) ion batteries, ACS Appl. Mater. Interf. 6 (2014) 11173-11179. https://doi.org/10.1021/am501144q

[37] H.M. Jeong, J.W. Lee, W.H. Shin, Y.J. Choi, H.J. Shin, J.K. Kang, J.W. Choi, Nitrogen-doped graphene for high-performance ultracapacitors and the importance of nitrogen-doped sites at basal planes, Nano Lett. 11 (2011) 2472-2477. https://doi.org/10.1021/nl2009058

[38] Z.H. Wen, X.C. Wang, S. Mao, Z. Bo, H. Kim, S.M. Cui, G.H. Lu, X.L. Feng, J.H. Chen, Crumpled nitrogen-doped graphene nanosheets with ultrahigh pore volume for high-performance supercapacitor, Adv. Mater. 24 (2012) 5610-5616. https://doi.org/10.1002/adma.201201920

[39] Y. Zhao, C.G. Hu, Y. Hu, H.H. Cheng, G.Q. Shi, L.T. Qu, A versatile, ultralight, nitrogen-doped graphene framework, Angew. Chem. Int. Edit. 51 (2012) 11371-11375. https://doi.org/10.1002/anie.201206554

[40] L. Hao, X.L. Li, L.J. Zhi, Carbonaceous electrode materials for supercapacitors, Adv. Mater. 25 (2013) 3899-3904. https://doi.org/10.1002/adma.201301204

[41] Y.Q. Zou, I.A. Kinloch, R.A.W. Dryfe, Nitrogen-doped and crumpled graphene sheets with improved supercapacitance, J. Mater. Chem. A 2 (2014) 19495-19499. https://doi.org/10.1039/c4ta04076g

[42] S. Livraghi, M.C. Paganini, E. Giamello, A. Selloni, C. Di Valentin, G. Pacchioni, Origin of photoactivity of nitrogen-doped titanium dioxide under visible light, J. Am. Chem. Soc. 128 (2006) 15666-15671. https://doi.org/10.1021/ja064164c

[43] R. Asahi, T. Morikawa, H. Irie, T. Ohwaki, Nitrogen-doped titanium dioxide as visible-light-sensitive photocatalyst: designs, developments, and prospects, Chem. Rev. 114 (2014) 9824-9852. https://doi.org/10.1021/cr5000738

[44] S. Sakthivel, H. Kisch, Photocatalytic and photoelectrochemical properties of nitrogen-doped titanium dioxide, Chemphyschem 4 (2003) 487-490. https://doi.org/10.1002/cphc.200200554

[45] Y.Y. Wen, T.E. Rufford, X.Z. Chen, N. Li, M.Q. Lyu, L.M. Dai, L.Z. Wang, Nitrogen-doped $Ti_3C_2T_x$ MXene electrodes for high-performance supercapacitors, Nano Energy 38 (2017) 368-376. https://doi.org/10.1016/j.nanoen.2017.06.009

[46] W.Z. Bao, L. Liu, C.Y. Wang, S. Choi, D. Wang, G.X. Wang, Facile synthesis of crumpled nitrogen-doped mxene nanosheets as a new sulfur host for lithium-sulfur batteries, Adv. Energy Mater. 8 (2018) 1702485-1702496. https://doi.org/10.1002/aenm.201702485

[47] M.Q. Zhao, C.E. Ren, Z. Ling, M.R. Lukatskaya, C.F. Zhang, K.L. Van Aken, M.W. Barsoum, Y. Gogotsi, Flexible MXene/carbon nanotube composite paper with high volumetric capacitance, Adv. Mater. 27 (2015) 339-345. https://doi.org/10.1002/adma.201404140

[48] Z. Ling, C.E. Ren, M.Q. Zhao, J. Yang, J.M. Giammarco, J.S. Qiu, M.W. Barsoum, Y. Gogotsi, Flexible and conductive MXene films and nanocomposites with high capacitance, P. Natl. Acad. Sci. USA 111 (2014) 16676-16681. https://doi.org/10.1073/pnas.1414215111

[49] E.A. Mayerberger, O. Urbanek, R.M. McDaniel, R.M. Street, M.W. Barsoum, C.L. Schauer, Preparation and characterization of polymer-$Ti_3C_2T_x$ (MXene) composite nanofibers produced via electrospinning, J. Appl. Polym. Sci. 134 (2017) 45295-45302. https://doi.org/10.1002/app.45295

[50] G.D. Zou, Z.W. Zhang, J.X. Guo, B.Z. Liu, Q.R. Zhang, C. Fernandez, Q.M. Peng, Synthesis of MXene/Ag composites for extraordinary long cycle lifetime lithium storage at high rates, ACS Appl. Mater. Inter. 8 (2016) 22280-22286. https://doi.org/10.1021/acsami.6b08089

[51] J.F. Zhu, Y. Tang, C.H. Yang, F. Wang, M.J. Cao, Composites of TiO_2 nanoparticles deposited on Ti_3C_2 MXene nanosheets with enhanced electrochemical performance, J. Electrochem. Soc. 163 (2016) A785-A791. https://doi.org/10.1149/2.0981605jes

[52] B. Ahmed, D.H. Anjum, Y. Gogotsi, H.N. Alshareef, Atomic layer deposition of SnO_2 on MXene for Li-ion battery anodes, Nano Energy 34 (2017) 249-256. https://doi.org/10.1016/j.nanoen.2017.02.043

[53] I.L. Medintz, H.T. Uyeda, E.R. Goldman, H. Mattoussi, Quantum dot bioconjugates for imaging, labelling and sensing, Nat. Mater. 4 (2005) 435-446. https://doi.org/10.1038/nmat1390

[54] D.Y. Pan, J.C. Zhang, Z. Li, M.H. Wu, Hydrothermal route for cutting graphene sheets into blue-luminescent graphene quantum dots, Adv. Mater. 22 (2010) 734-738. https://doi.org/10.1002/adma.200902825

[55] Q. Xue, H.J. Zhang, M.S. Zhu, Z.X. Pei, H.F. Li, Z.F. Wang, Y. Huang, Y. Huang, Q.H. Deng, J. Zhou, S.Y. Du, Q. Huang, C.Y. Zhi, Photoluminescent Ti_3C_2 MXene quantum dots for multicolor cellular imaging, Adv. Mater. 29 (2017) 1604847-1604853. https://doi.org/10.1002/adma.201604847

[56] X.H. Yu, X.K. Cai, H.D. Cui, S.W. Lee, X.F. Yu, B.L. Liu, Fluorine-free preparation of titanium carbide MXene quantum dots with high near-infrared photothermal performances for cancer therapy, Nanoscale 9 (2017) 17859-17864. https://doi.org/10.1039/c7nr05997c

[57] M. Khazaei, M. Arai, T. Sasaki, M. Estili, Y. Sakka, Two-dimensional molybdenum carbides: potential thermoelectric materials of the MXene family, Phys. Chem. Chem. Phys. 16 (2014) 7841-7849. https://doi.org/10.1039/c4cp00467a

[58] S. Wang, J.X. Li, Y.L. Du, C. Cui, First-principles study on structural, electronic and elastic properties of graphene-like hexagonal Ti_2C monolayer, Comp. Mater. Sci. 83 (2014) 290-293. https://doi.org/10.1016/j.commatsci.2013.11.025

[59] H. Lashgari, M.R. Abolhassani, A. Boochani, S.M. Elahi, J. Khodadadi, Electronic and optical properties of 2D graphene-like compounds titanium carbides and nitrides: DFT calculations, Solid State Commun. 195 (2014) 61-69. https://doi.org/10.1016/j.ssc.2014.06.008

[60] M. Khazaei, M. Arai, T. Sasaki, M. Estili, Y. Sakka, The effect of the interlayer element on the exfoliation of layered Mo(2)AC (A = Al, Si, P, Ga, Ge, As or In) MAX phases into two-dimensional Mo_2C nanosheets, Sci. Technol. Adv. Mater. 15 (2014) 014208-014215. https://doi.org/10.1088/1468-6996/15/1/014208

[61] X.Z. Chen, Z.Z. Kong, N. Li, X.J. Zhao, C.H. Sun, Proposing the prospects of Ti3CN transition metal carbides (MXenes) as anodes of Li-ion batteries: a DFT study, Phys. Chem. Chem. Phys. 18 (2016) 32937-32943. https://doi.org/10.1039/c6cp06018h

[62] Q.Q. Meng, J.L. Ma, Y.H. Zhang, Z. Li, C.Y. Zhi, A. Hu, J. Fan, The S-functionalized Ti3C2 Mxene as a high capacity electrode material for Na-ion batteries: a DFT study, Nanoscale 10 (2018) 3385-3392. https://doi.org/10.1039/c7nr07649e

[63] X. Liang, Y. Rangom, C.Y. Kwok, Q. Pang, L.F. Nazar, Interwoven mxene nanosheet/carbon-nanotube composites as Li-S cathode hosts, Adv. Mater. 29 (2017) 1603040-1603047. https://doi.org/10.1002/adma.201603040

[64] M. Kurtoglu, M. Naguib, Y. Gogotsi, M.W. Barsoum, First principles study of two-dimensional early transition metal carbides, MRS Commun. 2 (2012) 133-137. https://doi.org/10.1557/mrc.2012.25

[65] [64] M.R. Lukatskaya, O. Mashtalir, C.E. Ren, Y. Dall'Agnese, P. Rozier, P.L. Taberna, M. Naguib, P. Simon, M.W. Barsoum, Y. Gogotsi, Cation intercalation and high volumetric capacitance of two-dimensional titanium carbide, Science 341 (2013) 1502-1505. https://doi.org/10.1126/science.1241488

[66] M.M. Hu, Z.J. Li, H. Zhang, T. Hu, C. Zhang, Z. Wu, X.H. Wang, Self-assembled Ti3C2Tx MXene film with high gravimetric capacitance, Chem. Commun. 51 (2015) 13531-13533. https://doi.org/10.1039/c5cc04722f

[67] K. Fic, G. Lota, M. Meller, E. Frackowiak, Novel insight into neutral medium as electrolyte for high-voltage supercapacitors, Energy Environ. Sci. 5 (2012) 5842-5850. https://doi.org/10.1039/c1ee02262h

[68] J.K. Kim, E. Lee, H. Kim, C. Johnson, J. Cho, Y. Kim, Rechargeable seawater battery and its electrochemical mechanism, Chemelectrochem 2 (2015) 328-332. https://doi.org/10.1002/celc.201402344

[69] Q.X. Xia, N.M. Shinde, T.F. Zhang, J.M. Yun, A.G. Zhou, R.S. Mane, S. Mathur, K.H. Kim, Seawater electrolyte-mediated high volumetric MXene-based electrochemical symmetric supercapacitors, Dalton T. 47 (2018) 8676-8682. https://doi.org/10.1039/c8dt01375f

[70] X. Guo, X. Xie, S. Choi, Y. Zhao, H. Liu, C. Wang, S. Chang, G. Wang, Sb2O3/MXene (Ti3C2Tx) hybrid anode materials with enhanced performance for sodium-ion batteries, J. Mater. Chem. A 5 (2017) 12445-12452. https://doi.org/10.1039/c7ta02689g

[71] B.H. Dang, M. Rahman, D. MacElroy, D.P. Dowling, Evaluation of microwave plasma oxidation treatments for the fabrication of photoactive un-doped and carbon-doped TiO2 coatings, Surf. Coat. Tech. 206 (2012) 4113-4118. https://doi.org/10.1016/j.surfcoat.2012.04.003

[72] M. Hassan, R. Rawat, P. Lee, S. Hassan, A. Qayyum, R. Ahmad, G. Murtaza, M. Zakaullah, Synthesis of nanocrystalline multiphase titanium oxycarbide (TiC_xO_y) thin films by UNU/ICTP and NX_2 plasma focus devices, Appl. Phys. A 90 (2008) 669-677. https://doi.org/10.1007/s00339-007-4335-8

[73] J. Halim, K.M. Cook, M. Naguib, P. Eklund, Y. Gogotsi, J. Rosen, M.W. Barsoum, X-ray photoelectron spectroscopy of select multi-layered transition metal carbides (MXenes), Appl. Surf. Sci. 362 (2016) 406-417. https://doi.org/10.1016/j.apsusc.2015.11.089

[74] W.S. Epling, G.B. Hoflund, J.F. Weaver, S. Tsubota, M. Haruta, Surface characterization study of $Au/\alpha\text{-}Fe_2O_3$ and Au/Co_3O_4 low-temperature CO oxidation catalysts, J. Phys. Chem. 100 (1996) 9929-9934. https://doi.org/10.1021/jp960593t

[75] G. Liu, C. Han, M. Pelaez, D. Zhu, S. Liao, V. Likodimos, N. Ioannidis, A.G. Kontos, P. Falaras, P.S. Dunlop, Synthesis, characterization and photocatalytic evaluation of visible light activated C-doped TiO_2 nanoparticles, Nanotechnology 23 (2012) 294003-294013. https://doi.org/10.1088/0957-4484/23/29/294003

[76] G. Wang, Z. Ma, G. Zhang, C. Li, G. Shao, Cerium-doped porous K-birnessite manganese oxides microspheres as pseudocapacitor electrode material with improved electrochemical capacitance, Electrochim Acta 182 (2015) 1070-1077. https://doi.org/10.1016/j.electacta.2015.10.028

[77] M.F. El-Kady, R.B. Kaner, Scalable fabrication of high-power graphene micro-supercapacitors for flexible and on-chip energy storage, Nat. Commun. 4 (2013) 1475-1484. https://doi.org/10.1038/ncomms2446

[78] W. Gao, N. Singh, L. Song, Z. Liu, A.L.M. Reddy, L. Ci, R. Vajtai, Q. Zhang, B. Wei, P.M. Ajayan, Direct laser writing of micro-supercapacitors on hydrated graphite oxide films, Nat. Nanotechnol. 6 (2011) 496-500. https://doi.org/10.1038/nnano.2011.110

[79] C. Ogata, R. Kurogi, K. Awaya, K. Hatakeyama, T. Taniguchi, M. Koinuma, Y. Matsumoto, All-graphene oxide flexible solid-state supercapacitors with enhanced electrochemical performance, ACS Appl. Mater. Interf. 9 (2017) 26151-26160. https://doi.org/10.1021/acsami.7b04180

[80] D. Yu, K. Goh, Q. Zhang, L. Wei, H. Wang, W. Jiang, Y. Chen, Controlled functionalization of carbonaceous fibers for asymmetric solid-state micro-supercapacitors with high volumetric energy density, Adv. Mater. 26 (2014) 6790-6797. https://doi.org/10.1002/adma.201403061

[81] A.N. Enyashin, A.L. Ivanovskii, Two-dimensional titanium carbonitrides and their hydroxylated derivatives: Structural, electronic propertis and stability of MXenes $Ti_3C_{2-x}N_x(OH)_2$ from DFTB calculations, J. Solid State. Chem. 207 (2013) 42-48. https://doi.org/10.1016/j.jssc.2013.09.010

[82] Z.F. Lin, D. Barbara, P.L. Taberna, K.L. Van Aken, B. Anasori, Y. Gogotsi, P. Simon, Capacitance of $Ti_3C_2T_x$ MXene in ionic liquid electrolyte, J. Power Sources 326 (2016) 575-579. https://doi.org/10.1016/j.jpowsour.2016.04.035

[83] Q. Tang, Z. Zhou, P.W. Shen, Are MXenes promising anode materials for Li ion batteries? computational studies on electronic properties and Li storage capability of Ti_3C_2 and $Ti_3C_2X_2$ (X = F, OH) monolayer, J. Am. Chem. Soc. 134 (2012) 16909-16916. https://doi.org/10.1021/ja308463r

[84] R.B. Rakhi, B. Ahmed, M.N. Hedhili, D.H. Anjum, H.N. Alshareef, Effect of postetch annealing gas composition on the structural and electrochemical properties of Ti_2CT_x MXene electrodes for supercapacitor applications, Chem. Mater. 27 (2015) 5314-5323. https://doi.org/10.1021/acs.chemmater.5b01623

[85] M. Ghidiu, M.R. Lukatskaya, M.Q. Zhao, Y. Gogotsi, M.W. Barsoum, Conductive two-dimensional titanium carbide 'clay' with high volumetric capacitance, Nature 516 (2014) 78-U171. https://doi.org/10.1038/nature13970

[86] Y. Dall'Agnese, P.L. Taberna, Y. Gogotsi, P. Simon, Two-dimensional vanadium carbide (MXene) as positive electrode for sodium-ion capacitors, J. Phys. Chem. Lett. 6 (2015) 2305-2309. https://doi.org/10.1021/acs.jpclett.5b00868

[87] A. VahidMohammadi, A. Hadjikhani, S. Shahbazmohamadi, M. Beidaghi, Two-dimensional vanadium carbide (MXene) as a high-capacity cathode material for rechargeable aluminum batteries, ACS Nano 11 (2017) 11135-11144. https://doi.org/10.1021/acsnano.7b05350

[88] Q.M. Shan, X.P. Mu, M. Alhabeb, C.E. Shuck, D. Pang, X. Zhao, X.F. Chu, Y. Wei, F. Du, G. Chen, Y. Gogotsi, Y. Gao, Y. Dall'Agnese, Two-dimensional vanadium carbide (V_2C) MXene as electrode for supercapacitors with aqueous electrolytes, Electrochem. Commun. 96 (2018) 103-107. https://doi.org/10.1016/j.elecom.2018.10.012

[89] C.H. Yang, W.X. Que, X.T. Yin, Y.P. Tian, Y.W. Yang, M.D. Que, Improved capacitance of nitrogen-doped delaminated two-dimensional titanium carbide by urea-assisted synthesis, Electrochim. Acta 225 (2017) 416-424. https://doi.org/10.1016/j.electacta.2016.12.173

[90] Y.P. Tian, C.H. Yang, W.X. Que, X.B. Liu, X.T. Yin, L.B. Kong, Flexible and free-standing 2D titanium carbide film decorated with manganese oxide nanoparticles as a high volumetric capacity electrode for supercapacitor, J. Power Sources 359 (2017) 332-339. https://doi.org/10.1016/j.jpowsour.2017.05.081

[91] Y. Wang, H. Dou, J. Wang, B. Ding, Y.L. Xu, Z. Chang, X.D. Hao, Three-dimensional porous MXene/layered double hydroxide composite for high performance supercapacitors, J. Power Sources 327 (2016) 221-228. https://doi.org/10.1016/j.jpowsour.2016.07.062

[92] R.Z. Zhao, M.Q. Wang, D.Y. Zhao, H. Li, C.X. Wang, L.W. Yin, Molecular-level heterostructures assembled from titanium carbide MXene and Ni-Co-Al layered double-hydroxide nanosheets for all-solid-state flexible asymmetric high-energy supercapacitors, ACS Energy Lett. 3 (2018) 132-140. https://doi.org/10.1021/acsenergylett.7b01063

[93] D. Qu, L. Wang, D. Zheng, L. Xiao, B. Deng, D. Qu, An asymmetric supercapacitor with highly dispersed nano-Bi_2O_3 and active carbon electrodes, J. Power Sources 269 (2014) 129-135. https://doi.org/10.1016/j.jpowsour.2014.06.084

[94] J. Li, Q. Wu, G. Zan, A high-performance supercapacitor with well-dispersed Bi_2O_3 nanospheres and active-carbon electrodes, Eur. J. Inorg. Chem. 2015 (2015) 5751-5756. https://doi.org/10.1002/ejic.201500904

[95] Q.X. Xia, N.M. Shinde, J.M. Yun, T. Zhang, R.S. Mane, S. Mathur, K.H. Kim, Bismuth oxychloride/MXene symmetric supercapacitor with high volumetric energy density, Electrochim. Acta 271 (2018) 351-360. https://doi.org/10.1016/j.electacta.2018.03.168

[96] Z. Ling, C.E. Ren, M.-Q. Zhao, J. Yang, J.M. Giammarco, J. Qiu, M.W. Barsoum, Y. Gogotsi, Flexible and conductive MXene films and nanocomposites with high capacitance, P. Natl A. Sci. 111 (2014) 16676-16681. https://doi.org/10.1073/pnas.1414215111

[97] T.K. Zhao, J.K. Zhang, Z. Du, Y.H. Liu, G.L. Zhou, J.T. Wang, Dopamine-derived N-doped carbon decorated titanium carbide composite for enhanced supercapacitive performance, Electrochim. Acta 254 (2017) 308-319. https://doi.org/10.1016/j.electacta.2017.09.144

[98] R. Rakhi, B. Ahmed, M.N. Hedhili, D.H. Anjum, H.N. Alshareef, Effect of postetch annealing gas composition on the structural and electrochemical properties of Ti_2CT_x MXene electrodes for supercapacitor applications, Chem. Mater. 27 (2015) 5314-5323. https://doi.org/10.1021/acs.chemmater.5b01623

MXenes: Fundamentals and Applications   Materials Research Forum LLC
Materials Research Foundations **51** (2019) 137-174   doi: https://doi.org/10.21741/9781644900253-6

[99] R.B. Rakhi, B. Ahmed, D. Anjum, H.N. Alshareef, Direct chemical synthesis of MnO_2 nanowhiskers on transition-metal carbide surfaces for supercapacitor applications, ACS Appl. Mater. Interf. 8 (2016) 18806-18814. https://doi.org/10.1021/acsami.6b04481

[100] A. Jagadale, V. Kumbhar, D. Dhawale, C. Lokhande, Performance evaluation of symmetric supercapacitor based on cobalt hydroxide $[Co(OH)_2]$ thin film electrodes, Electrochim. Acta 98 (2013) 32-38. https://doi.org/10.1016/j.electacta.2013.02.094

[101] Y. He, W. Chen, X. Li, Z. Zhang, J. Fu, C. Zhao, E. Xie, Freestanding three-dimensional graphene/MnO_2 composite networks as ultralight and flexible supercapacitor electrodes, ACS Nano 7 (2012) 174-182. https://doi.org/10.1021/nn304833s

[102] A. Chu, P. Braatz, Comparison of commercial supercapacitors and high-power lithium-ion batteries for power-assist applications in hybrid electric vehicles: I. Initial characterization, J. Power Sources 112 (2002) 236-246. https://doi.org/10.1016/s0378-7753(02)00364-6

[103] B. Anasori, M.R. Lukatskaya, Y. Gogotsi, 2D metal carbides and nitrides (MXenes) for energy storage, Nat. Rev. Mater. 2 (2017) 16098-16115. https://doi.org/10.1038/natrevmats.2016.98

MXenes: Fundamentals and Applications   Materials Research Forum LLC
Materials Research Foundations **51** (2019) 175-188   doi: https://doi.org/10.21741/9781644900253-7

Chapter 7

MXenes for Sodium-Ion Batteries

Rashid Iqbal[1,2], Muhammad Qaisar Sultan[1,2], Ramyakrishna Pothu[3], Rajender Boddula[1]*

[1] CAS Key Laboratory of Nanosystem and Hierarchical Fabrication, CAS Center for Excellence in Nanoscience, National Center for Nanoscience and Technology, Beijing 100190, China

[2] University of Chinese Academy of Sciences, Beijing 100039, China

[3] College of Chemistry and Chemical Engineering, Hunan University, Changsha 410082, China

B.Rajender, research.raaj@qq.com

Abstract

MXene is a new class of porous two-dimensional materials that have attracted enormous attention during the last decades due to their high redox activity controllable structures and tunable pore sizes. In this review, we report some recent advances in titanium carbide based MXene based electrode used without binder, conductive additive and current collector in the fabrication of next-generation batteries. This MXene electrode is proficient of redox reversible electrochemical storage of sodium ions and possesses good cycling stability with high rate charge-discharge ability. Meanwhile, Li metal is a scarce and expensive element in the earth crust and may not be able to resolve the energy requirement in the future, whereas Na metal could easily replace Li metal due to its readily availability. Finally, this review will focus on the research, which will open up new opportunities for developing self-standing binder and additive-free MXene electrodes for Na-ion batteries with energy density approaching those of Li-ion.

Keywords

MXene, Two-Dimensional Material, Non-Lithium-Ion Batteries, Sodium Ion Batteries

Contents

1.   Introduction

During the last two decades, rechargeable batteries have industrialized and are applicable in daily used electronic appliances (electric vehicle, portable energy storage devices) in today's modern society. The storage of energy from renewable energy sources like solar cell, wind turbines is essential. Consequently, rechargeable batteries are needed to combat climate change, which is a global issue these days. Two-dimensional (2D) materials have a variety of advantage to be used in the field of rechargeable batteries due to high surface area, which is absent in the bulky conventional materials. MXenes are a new class of 2D materials, first reported in 2011 [1, 2]. MXene are made up of carbon nitride and transition metals with general formula of $M_{n+1}X_nT_x$, metal is represented by M, X is for carbon and/or nitrogen, where n represents the number (1, 2, 3, …) and T_x stands for the surface functional groups which could be (F, O and OH) [3-6]. So far MXene have been used in numerous different energy storage devices like Na-ion, Li-ion, Li-S and supercapacitors. MXenes have revealed potential for different applications and especially for the energy storage in Li-ion [7], Li-S [8-10], Na-ion batteries [3-6], and supercapacitors as shown in (Fig. 1) [11-13].

2D materials have the inherited problem of stacking, consequently hinders the ionic transport within the two electrodes, this behavior is consistent in MXene. To take full advantage of MXene for achieving better electrochemical performance by synthesizing porous nanostructures and insertion of foreign species or functional groups in between MXene layers. Mesoporous MXene performs excellent performance with increased Li-storage by factor of four along with excellent rate performance [14]. DFT calculation in ab initio mode did pioneering work in the study of MXene to determine enlargement in the interstitial space of $Ti_3C_2T_x$ by nano-carbon material [15, 16], polymeric material [17, 18], metallic nanoparticles [19], and enormous ions [20] allows adsorption of numerous layer of sodium-ion (Na-ion), thus have potential for high capacity batteries [21]. Therefore, as lithium is an expensive metal with less abundance, urgent need exist to explore various abundant metal-ion batteries like Na-ion [22, 23], Mg-ion [24-27], Al-ion [28, 29], K-ion [30, 31] and Ca-ion [32, 33]. These are appealing for fulfilling high theoretical capacities demand with appropriate negative redox functionality, working protection, and are environment friendly. As it is well known that the electrochemical performance of batteries depends upon the properties of the electrode and electrolyte materials. The research and development of anode materials is quite sluggish and restricted to carbon related materials as compared to the fast progress in cathode materials, where group IVA and VA centered metal alloys that are capable anode materials for LIBs [34-41]. However, most of these materials are inappropriate to be utilize for NLIBs. For example, graphite, the furthermost frequently used anode material

MXenes: Fundamentals and Applications
Materials Research Foundations 51 (2019) 175-188

Materials Research Forum LLC
doi: https://doi.org/10.21741/9781644900253-7

in LIBs, displays almost no electrochemical performance for NLIBs [42]. Furthermore, among other variety of carbon materials [34-36] elucidate good cyclic performance at low charge-discharge rates with a mediocre capacity. Conversely, metal alloys can displays high capacities with disadvantage of huge volume increase inside the battery [37,41] which ultimately becomes the origin for a breakdown of the electrode. Consequently, the research and improvement in performance as an anode material with reversible redox reaction, rapid ionic movement within the electrode and high rate charge-discharge with exceptional cyclic capability is a foremost issue for NLIBs.

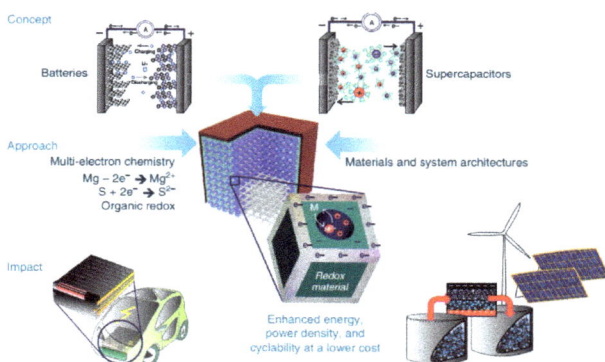

Figure 1.Schematic representation of rechargeable batteries and various energy storage devices based on MXene [Reprinted with permission from ref. 13. Copyright 2016 Nat. Comm., Nature].

2.   Na-ion batteries

More precisely, after commercialization of lithium-ion (LIBs) batteries by Sony in 1991, they received a lot of attention because of their broad voltage range and high energy density. LIBs possess numerous challenges, which hinders its futuristic practical application due to cost, safety and the recently used material reached their theoretical capacity [43]. Hence, the batteries required other metal ion to possess higher theoretical capacity. LIBs use is growing in electric vehicles and its large level storage in energy grid system, thus lithium is insufficient for future batteries as it could not be able to meet the energy supply chain. Furthermore, as sodium metal is available in plentiful amount in earth and is a cheaper material as compare to lithium metal. With future innovative research methodologies Na-ion batteries (NIBs) have potential to easily replace LIBs [23,44]. In addition, numerous other metal ions could be used like potassium, multivalent ions, for example, aluminum, calcium and magnesium could provide higher energy density as compare to monovalent Li, based on two three electron transfer per Li-ion [45]. Although, these new metal ions explained above needs to be researched and developed to realize the practical applications. Finding a appropriate electrolyte and suitable redox active host material is required as graphite cannot be used in the Na-ion batteries, where Na-C bonding is very weak as compare to Li-ion batteries [46].

Most recently reported studies for non-Lithium-ion batteries (NLIBs) are investigated to develop cathode materials. Numerous 2D materials have been synthesized with large interstitial spaces distance, with layered transition-metal oxides [12], covalent framework materials [14], and olivine have been examined as cathodes displayed good capacities and better cyclic performance.

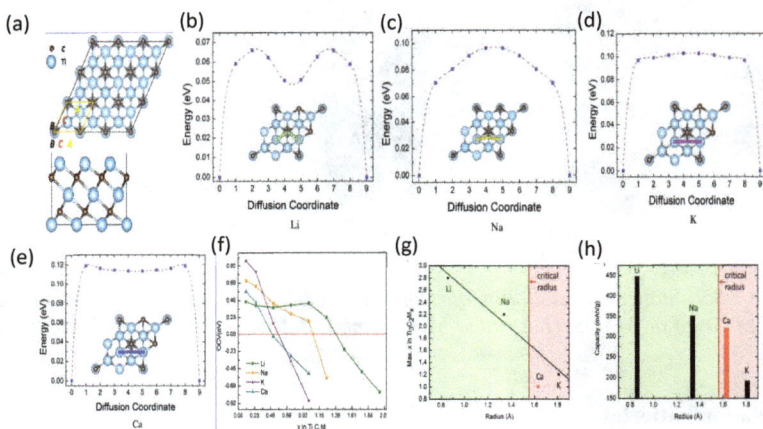

Figure 2. Schematic diagram demonstrates the crystal of Ti₃C₂ monolayer with top and side view. Blue color represents Ti atoms and brown indicates Carbon atoms. parallel diffusion barrier profiles of (b) Li, (c) Na, (d) K, and (e) Ca on Ti₃C₂ MXene. OCV changes with adatom quantity for the single-side adsorption of Li, Na, K, and Ca on the Ti₃C₂surface [Reprinted with permission from ref. 47. Copyright 2014ACS Appl. Mater. Interfaces, ACS].

A new kind of MXene constructed by eliminating the Al atom from bulk Ti₃AlC₂, to obtain the monolayer with quintuple layers packed in an order of Ti−C−Ti(c)−C−Ti, where Ti(c) stands for central Ti atoms in Ti₃C₂ [47]. The tranquil structure was calculated theoretically to have a lattice constant of a = 3.1005 Å, which is consistent

Materials Research Forum LLC
doi: https://doi.org/10.21741/9781644900253-7

with the experimental calculated lattice constant of 3.057 Å [48]. This could be also explained like this that trilayered material of Ti sandwiched the two Carbon atomic layers formed TiC_6 (octahedral structure) in Fig. 2. Highly symmetric atomic sites A, B, and C are easy to recognize due to the adsorption on the surface of the material as displayed in Fig. 2. Central position of hexagon is occupied by the A site, which is occupied by carbon atoms, where the B site is overhead the A and C site is occupied by the Ti atoms. Furthermore, the interstitial space has less probability of metal atoms as it is less energetically favorable as compare to surface adsorption. According to the theoretical calculations the adoption energies of metal atoms in interstitial space like, Na, Ca Li and K showed 12.23, 12.21, 4.40, 15.39 eV, disturbed the structure up to some extent, correspondingly. As previously well studied the diffusion barrier is responsible for the charging-discharging rate of the battery. In this study, diffusion barriers for the various metal-ions were investigated in the MXene monolayer named Na, Li, K, and causing Vienna Ab Initio Simulation Package (VASP) in nudged elastic band technique to determine the electrochemical performance of this 2D material. The movement of ions were tuned among highly probable and energetically feasible adoption position on the surface of the material (Fig. 2f) [49]. The most favorable diffusion pathway for Lithium is calculated to be 0.068 eV, which is similar to previous reports. By using the similar theoretical calculation model K, Na and Ca diffusion barrier were intended to be 0.103, 0.096, and 0.118 eV, correspondingly [50]. Like TiO_2 polymorphs commercially used anode material have low diffusion barrier for Li, as compare MXene Ti_3C_2 displayed faster Li-ion movement and higher rate performance also showed promising results with K, Na and Ca. The open circuit voltage (OCV) demonstrates the adatom quantity as function of X in the type of MXene system ($Ti_3C_2T_x$). The different concentration of alkali atoms demonstrated OCV profile with Li have lowest property, while property decreases with increased adatom quantity having various slopes. Furthermore, other alkali metals OCV reductions as the x value upsurge, where Li-ion displays small variation with change value of x is less than 1. The gap between OCV of A and B site is around 100 millivolts suggesting the unique precise tendency. High adatom increases OCV among alkali metals and opposite trend by decreasing the x value of adatom quantity. The calculated value for K and Ca for is x is 0.6 and 0.5, where no further adatoms could be adsorbed, correspondingly, OCV reduced to zero value before completely shielding A. OCV is positive when x is equal to 1 demonstrates that more adatoms could be adsorb on the surface of MXene, meanwhile A is completely shielded. Furthermore, the metal ions of Na and Li possess the extreme x value of 1.1 and 1.4, correspondingly, explaining the directly proportional association among the both adatoms close to zero value of OCV. These investigations give information about the covering of interstitial space between the layers which energetically most promising site to adsorb more adatoms afterward completely shielding site A.

Density functional theory (DFT) technique using along with the first principle examined the adsorption properties of alkali metals used in batteries on MXene. The investigation suggested that the direct relationship of effective radius size of adatom, its content and also ability to charge-discharge. Charge movement depends upon the radii of the metal ion and thus it is a very crucial parameter for adsorption of alkali metal on the surface of MXene. Bigger size metal ion increases the bonding interaction between the alkali metal ions, therefore, reduces the quantity of adatom and suggesting deprived theoretical specific capacity of the battery. The capacitance values for the various alkali metal ion batteries are calculated to be 351.8, 319.8, 447.8 and 191.8 mAh g^{-1} for Na, Ca, Li, and K-ion batteries, correspondingly. These investigations provide the deep understanding for futuristic research in the field of MXene with varied alkali metal-ion batteries.

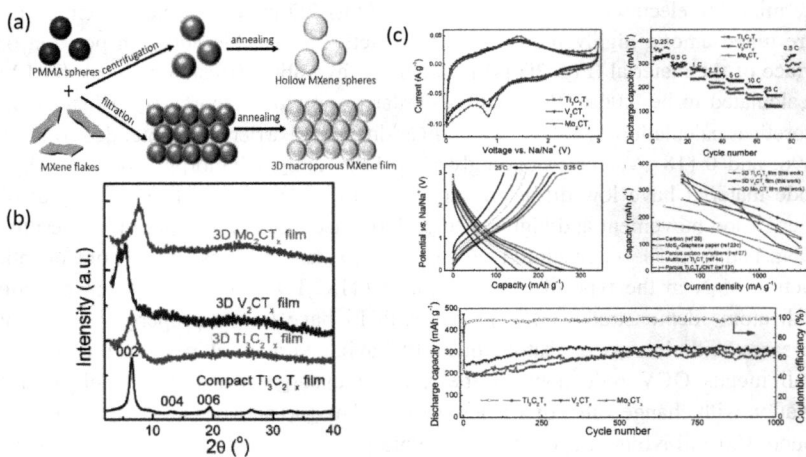

Figure 3. (a). Schematic diagram demonstrates the synthesis of hollow MXene spheres and 3D macroporous MXene frameworks also demonstrates the removal of PMMA by heat treatment at 450 °C. (b). XRD profiles of 3D macroporous MXene films. (c) Electrochemical performances of the 3D macroporous MXene film electrodes for Na-ion storage [Reprinted with permission from ref. 51. Copyright 2017Adv. Mater., Wiley].

In another study, researcher have demonstrated the synthesis of 3D microporous MXene from the composited of PMMA and MXene flakes as presented in (Fig. 3) [51] rigid interaction were formed between PMMA and MXene flakes. MXene fakes covered the PMMA nanospheres completely bonded by the hydroxide functional groups. Thermal treatment at 450 °C and followed by filtration is used to separate the PMMA and MXene, results in removal of PMMA, thus self-standing hollow MXene spheres could be achieved. Similarly, centrifugation followed by heat treatment leads to the formation of 3D self-standing macroporous MXene film after removal of PMMA. X-ray diffraction (XRD) demonstrates the profile of bulk $Ti_3C_2T_x$ flakes, its 3D macroporous films, V_2CT_x, and Mo_2CT_x films. Inherent (002) diffraction peaks for MXenes were practical found in all of the films. $Ti_3C_2T_x$ flakes displays loss of (001), (004) and (006), due to the presences of macroporous 3D flakes construction with haphazard alignment [14]. 3D MXene films were further characterized by cyclic voltammetry (CV) curves to investigate their electrochemical behavior as displayed in (Fig. 3c). Redox peaks located 0.8 and 1.4 V for Na-ion corresponds to movement through the electrode material. The bigger width of redox peaks indicates a reaction which is not diffusion limited and sodium ion storage mechanism is based on pseudocapacitance, analogous as reported before [3, 6, 52]. Cathodic peaks were experimentally displayed in the initial cycle, which is accredited to the development of a solid-state electrolyte boundary and the reaction of sodium ions with functional groups of fluorine, hydroxide, oxygen or water molecules resides between the flakes of MXene. In comparison of three different kinds of 3D MXenes, Mo_2CT_x demonstrates the capacity of 370 mAh g^{-1} at 0.25C, whereas V_2CT_x and $Ti_3C_2T_x$ gives capacity of 340 and 330 mAh g^{-1}, correspondingly. These films showed brilliant capacity retention, moreover, at 25 C V_2CT_x demonstrate the highest capacity of 170 mAh g^{-1}, due to the presence of bigger interlayer distance [53]. 3D $Ti_3C_2T_x$ film shows broad redox peaks at 0.25 C but these redox peaks vanished to higher charge-discharge rates confirm the mechanism of pseudocapacitive sodium-ion storage (Fig. 3c). Thus these MXene have the potential to perform for redox Na-ion Supercapacitors application. These MXene films shows promising reversible capacity after 1000 cycles with first cycle columbic efficiency higher than 50%, which is much higher as compare to formerly studies material $Ti_3C_2T_x$/CNT electrode, which is about 40% [6].

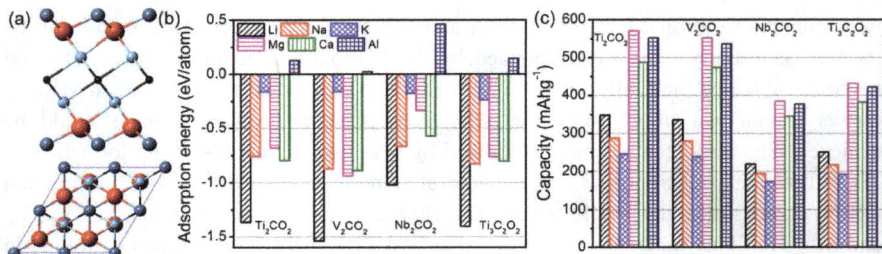

Figure 4.(a) Different angle schematic diagrams of Ti_2CO_2 nanosheet, (b) represents ion adsorption energies, (c) theoretical capacities of bare MXene nanosheets [Reprinted with permission from ref. 53. Copyright 2014ACS Nano, ACS].

In another study, the surface modification is really important and key method to interact with alkali metal-ions for storage purposes, e.g., hydroxide conversion to oxygen functional group upon annealing [47, 52, 54, 55]. The synthesis of bare MXene from terminated MXene is a new dimension to explore its potential for non-Lithium-ion batteries (NLIBs). In this study understanding of the mechanism of intercalation of various metal ion batteries like Na^+, K^+, Mg^{2+}, Ca^{2+}, and Al^{3+} on MXene nanosheets surface with the help of DFT and experimental electrochemical performances as shown in (Fig. 4) [53]. In specific, the Oxygen terminated shows unfavorable performance compare to bare MXene, which is proven to be excellent of alkali metal ion batteries. Na^+ and K^+ ion batteries perform exceptionally well by using bare Ti_3C_2 nanosheet, as proven by DFT calculations. On the other hand, storage mechanism for Al and Mg batteries is based on the adsorption of multilayer.

According to simulations results the MXenes storage mechanism of alkali metals reveals to be more convoluted as compare to previously reported anodic electrode material. In this study, initially MXenes with OH-terminated functional groups used acquire the O-termination MXene, further react to produce bare MXenes and metallization carried out in order to decorate metal oxides as planned alteration reactions. Metal ions were stored in the MXene as the reaction is reversible. Mostly these metal ions adsorbed layer in the interstitial space. Whereas, storing additional ions Mg and Al by formation of numerous metal layers could enhance the capacity further. Consequently, three different kinds of storage mechanism initiated in MXenes like plating/stripping, reversible conversion, and insertion/extraction. This MXene material possesses two major properties of high capacities and low diffusion barriers make materials for metal ion storage, predominantly, for alkali metal ions including Na-ion.

3. Summary

In conclusion, investigations of potential MXene electrode materials suggest the crucial requirement to determine appropriate electrolytes. The electrolyte is seemingly the important constituent to support NLIBs application of MXenes, which is yet to be explored. As the bare MXenes proved to be the more suitable material among terminated MXene since it demonstrates superior reactivity as compare to terminated MXenes because of the presence of dangling bonds on its surface. Due to the ease of solubility and synthesis MXenes are suitable to be produced in large scale as well as are environmentally friendly. MXene films could be used as free-standing material without use of binder and gives higher conductivity due to intrinsic property of this MXene. Furthermore, this material possesses the capability to be used without current collector as compare to conventional batteries, which is a big advantage for this material in Na-ion energy storage. MXenes makes it the futuristic material, which demonstrates excellent properties when composited with other carbon-based materials like CNTs or graphene. Therefore, in the future MXene is the promising material to be used in Na-ion energy storage with numerous advantages of environmentally benign and highly electrochemical active features.

Abbreviations

2D	Two-dimensional
Ca	Calcium
K	Potassium
Li	Lithium
LIBs	Lithium-ion batteries
NLIBs	Non-lithium-ion batteries
Na	Sodium
NIBs	Sodium-ion batteries
OCV	Open circuit voltage
VASP	Vienna ab initio simulation package

References

[1] B. Anasori, M.R. Lukatskaya, Y. Gogotsi, 2D metal carbides and nitrides (MXenes) for energy storage, Nat. Rev. Mater. 2 (2017) 16098. https://doi.org/10.1038/natrevmats.2016.98

[2] M. Naguib, V.N. Mochalin, M.W. Barsoum, Y. Gogotsi, 25th Anniversary Article: MXenes: A new family of two-dimensional materials, Adv. Mater. 26 (2014) 992-1005. https://doi.org/10.1002/adma.201304138

[3] X. Wang, S. Kajiyama, H. Iinuma, E. Hosono, S. Oro, I. Moriguchi, M. Okubo, A. Yamada, Pseudocapacitance of MXene nanosheets for high-power sodium-ion hybrid capacitors, Nat. Commun. 6 (2015) 6544. https://doi.org/10.1038/ncomms7544

[4] Y. Dall'Agnese, P.-L. Taberna, Y. Gogotsi, P. Simon, Two-dimensional vanadium carbide (MXene) as positive electrode for sodium-ion capacitors, J. Phys. Chem. Lett. 6 (2015) 2305-2309. https://doi.org/10.1021/acs.jpclett.5b00868

[5] M. Naguib, J. Halim, J. Lu, K.M. Cook, L. Hultman, Y. Gogotsi, M.W. Barsoum, New two-dimensional niobium and vanadium carbides as promising materials for Li-ion batteries, J. Am. Chem. Soc. 135 (2013) 15966-15969. https://doi.org/10.1021/ja405735d

[6] X. Xie, M.-Q. Zhao, B. Anasori, K. Maleski, C.E. Ren, J. Li, B.W. Byles, E. Pomerantseva, G. Wang, Y. Gogotsi, Porous heterostructured MXene/carbon nanotube composite paper with high volumetric capacity for sodium-based energy storage devices, Nano Energy 26 (2016) 513-523. https://doi.org/10.1016/j.nanoen.2016.06.005

[7] M. Naguib, O. Mashtalir, J. Carle, V. Presser, J. Lu, L. Hultman, Y. Gogotsi, M.W. Barsoum, Two-dimensional transition metal carbides, ACS Nano 6 (2012) 1322-1331. https://doi.org/10.1021/nn204153h

[8] X. Liang, A. Garsuch, L.F. Nazar, Sulfur cathodes based on conductive MXene nanosheets for high-performance lithium-sulfur batteries, Angew.Chem. Int. Ed. 54 (2015) 3907-3911. https://doi.org/10.1002/anie.201410174

[9] X. Liang, Y. Rangom, C.Y. Kwok, Q. Pang, L.F. Nazar, Interwoven MXene nanosheet/carbon nanotube composites as Li–S cathode hosts, Adv. Mater. 29 (2017) 1603040. https://doi.org/10.1002/adma.201603040

[10] X. Liang, A. Garsuch, L.F. Nazar, Sulfur cathodes based on conductive MXene nanosheets for high performance lithium-sulfur batteries, Angew.Chem. Int. Ed. 54 (2015) 3907-3911. https://doi.org/10.1002/anie.201410174

[11] M. Ghidiu, M.R. Lukatskaya, M.-Q. Zhao, Y. Gogotsi, M.W. Barsoum, Conductive two-dimensional titanium carbide 'clay' with high volumetric capacitance, Nature 516 (2014) 78. https://doi.org/10.1038/nature13970

[12] M.R. Lukatskaya, O. Mashtalir, C.E. Ren, Y. Dall'Agnese, P. Rozier, P.L. Taberna, M. Naguib, P. Simon, M.W. Barsoum, Y. Gogotsi, Cation intercalation and high volumetric capacitance of two-dimensional titanium carbide, Science 341 (2013) 1502-1505. https://doi.org/10.1126/science.1241488

[13] M.R. Lukatskaya, B. Dunn, Y. Gogotsi, Multidimensional materials and device architectures for future hybrid energy storage, Nat Commun. 7 (2016) 12647. https://doi.org/10.1038/ncomms12647

[14] C.E. Ren, M.Q. Zhao, T. Makaryan, J. Halim, M. Boota, S. Kota, B. Anasori, M.W. Barsoum, Y. Gogotsi, Porous Two-dimensional transition metal carbide (MXene) flakes for high performance Li ion storage, ChemElectroChem 3 (2016) 689-693. https://doi.org/10.1002/celc.201600059

[15] M.Q. Zhao, C.E. Ren, Z. Ling, M.R. Lukatskaya, C. Zhang, K.L. Van Aken, M.W. Barsoum, Y. Gogotsi, Flexible MXene/carbon nanotube composite paper with high volumetric capacitance, Adv. Mater. 27 (2015) 339-345. https://doi.org/10.1002/adma.201404140

[16] Y. Lu, L. Wang, J. Cheng, J.B. Goodenough, Prussian blue: a new framework of electrode materials for sodium batteries, Chem. Commun. 48 (2012) 6544-6546. https://doi.org/10.1039/c2cc31777j

[17] Z. Ling, C.E. Ren, M.-Q. Zhao, J. Yang, J.M. Giammarco, J. Qiu, M.W. Barsoum, Y. Gogotsi, Flexible and conductive MXene films and nanocomposites with high capacitance, PNAS 111 (2014) 16676-16681. https://doi.org/10.1073/pnas.1414215111

[18] M. Boota, B. Anasori, C. Voigt, M.Q. Zhao, M.W. Barsoum, Y. Gogotsi, Pseudocapacitive electrodes produced by oxidant free polymerization of pyrrole between the layers of 2D titanium carbide (MXene), Adv. Mater. 28 (2016) 1517-1522. https://doi.org/10.1002/adma.201504705

[19] M.-Q. Zhao, M. Torelli, C.E. Ren, M. Ghidiu, Z. Ling, B. Anasori, M.W. Barsoum, Y. Gogotsi, 2D titanium carbide and transition metal oxides hybrid electrodes for Li-ion storage, Nano Energy 30 (2016) 603-613. https://doi.org/10.1016/j.nanoen.2016.10.062

[20] J. Luo, W. Zhang, H. Yuan, C. Jin, L. Zhang, H. Huang, C. Liang, Y. Xia, J. Zhang, Y. Gan, Pillared structure design of MXene with ultralarge interlayer spacing for high-performance lithium-ion capacitors, ACS Nano 11 (2017) 2459-2469. https://doi.org/10.1021/acsnano.6b07668

[21] Y.-X. Yu, Prediction of mobility, enhanced storage capacity, and volume change during sodiation on interlayer-expanded functionalized Ti_3C_2 MXene anode materials for sodium-ion batteries, J. Phys. Chem. C 120 (2016) 5288-5296. https://doi.org/10.1021/acs.jpcc.5b10366

[22] V. Palomares, P. Serras, I. Villaluenga, K.B. Hueso, J. Carretero-González, T. Rojo, Na-ion batteries, recent advances and present challenges to become low-cost energy

storage systems, Energy Environ. Sci. 5 (2012) 5884-5901.
https://doi.org/10.1039/c2ee02781j

[23] M.D. Slater, D. Kim, E. Lee, C.S. Johnson, Sodium-ion batteries, Adv. Funct. Mater. 23 (2013) 947-958. https://doi.org/10.1002/adfm.201200691

[24] H.D. Yoo, I. Shterenberg, Y. Gofer, G. Gershinsky, N. Pour, D. Aurbach, Mg rechargeable batteries: an on-going challenge, Energy Environ. Sci. 6 (2013) 2265-2279. https://doi.org/10.1039/c3ee40871j

[25] D. Aurbach, Z. Lu, A. Schechter, Y. Gofer, H. Gizbar, R. Turgeman, Y. Cohen, M. Moshkovich, E. Levi, Prototype systems for rechargeable magnesium batteries, Nature 407 (2000) 724-727. https://doi.org/10.1038/35037553

[26] N. Amir, Y. Vestfrid, O. Chusid, Y. Gofer, D. Aurbach, Progress in nonaqueous magnesium electrochemistry, J Power Sour. 174 (2007) 1234-1240. https://doi.org/10.1016/j.jpowsour.2007.06.206

[27] R.E. Doe, R. Han, J. Hwang, A.J. Gmitter, I. Shterenberg, H.D. Yoo, N. Pour, D. Aurbach, Novel, electrolyte solutions comprising fully inorganic salts with high anodic stability for rechargeable magnesium batteries, Chem. Commun. 50 (2014) 243-245. https://doi.org/10.1039/c3cc47896c

[28] N. Jayaprakash, S.K. Das, L.A. Archer, The rechargeable aluminum-ion battery, Chem. Commun. 47 (2011) 12610-12612. https://doi.org/10.1039/c1cc15779e

[29] Q. Li, N.J. Bjerrum, Aluminum as anode for energy storage and conversion: a review, J. Power Sour. 110 (2002) 1-10.

[30] A. Eftekhari, Potassium secondary cell based on Prussian blue cathode, J. Power Sour. 126 (2004) 221-228. https://doi.org/10.1016/j.jpowsour.2003.08.007

[31] C.D. Wessells, R.A. Huggins, Y. Cui, Copper hexacyanoferrate battery electrodes with long cycle life and high power, Nat. Commun. 2 (2011) 550. https://doi.org/10.1038/ncomms1563

[32] G. Amatucci, F. Badway, A. Singhal, B. Beaudoin, G. Skandan, T. Bowmer, I. Plitz, N. Pereira, T. Chapman, R. Jaworski, Investigation of yttrium and polyvalent ion intercalation into nanocrystalline vanadium oxide, J. Electrochem. Soc. 148 (2001) A940-A950. https://doi.org/10.1149/1.1383777

[33] R.Y. Wang, C.D. Wessells, R.A. Huggins, Y. Cui, Highly reversible open framework nanoscale electrodes for divalent ion batteries, Nano Lett. 13 (2013) 5748-5752. https://doi.org/10.1021/nl403669a

[34] Y. Cao, L. Xiao, M.L. Sushko, W. Wang, B. Schwenzer, J. Xiao, Z. Nie, L.V. Saraf, Z. Yang, J. Liu, Sodium ion insertion in hollow carbon nanowires for battery applications, Nano Lett. 12 (2012) 3783-3787. https://doi.org/10.1021/nl3016957

[35] R. Alcántara, J.M. Jiménez-Mateos, P. Lavela, J.L. Tirado, Carbon black: a promising electrode material for sodium-ion batteries, Electrochem. Commun. 3 (2001) 639-642. https://doi.org/10.1016/s1388-2481(01)00244-2

[36] K. Tang, L. Fu, R.J. White, L. Yu, M.-M. Titirici, M. Antonietti, J. Maier, Hollow carbon nanospheres with superior rate capability for sodium-based batteries, Adv. Energy Mater. 2 (2012) 873-877. https://doi.org/10.1002/aenm.201100691

[37] A. Darwiche, C. Marino, M.T. Sougrati, B. Fraisse, L. Stievano, L. Monconduit, Better cycling performances of bulk Sb in Na-ion batteries compared to Li-ion systems: an unexpected electrochemical mechanism, J. Am. Chem. Soc. 134 (2012) 20805-20811. https://doi.org/10.1021/ja310347x

[38] Y. Zhu, X. Han, Y. Xu, Y. Liu, S. Zheng, K. Xu, L. Hu, C. Wang, Electrospun Sb/C fibers for a stable and fast sodium-ion battery anode, ACS Nano 7 (2013) 6378-6386. https://doi.org/10.1021/nn4025674

[39] J. Qian, Y. Xiong, Y. Cao, X. Ai, H. Yang, Synergistic Na-storage reactions in Sn_4P_3 as a high-capacity, cycle-stable anode of Na-ion batteries, Nano Lett. 14 (2014) 1865-1869. https://doi.org/10.1021/nl404637q

[40] Y. Shao, M. Gu, X. Li, Z. Nie, P. Zuo, G. Li, T. Liu, J. Xiao, Y. Cheng, C. Wang, J.-G. Zhang, J. Liu, Highly reversible Mg insertion in nanostructured Bi for Mg ion batteries, Nano Lett. 14 (2014) 255-260. https://doi.org/10.1021/nl403874y

[41] N. Singh, T.S. Arthur, C. Ling, M. Matsui, F. Mizuno, A high energy-density tin anode for rechargeable magnesium-ion batteries, Chem. Commun. 49 (2013) 149-151. https://doi.org/10.1039/c2cc34673g

[42] P. Ge, M. Fouletier, Electrochemical intercalation of sodium in graphite, Solid State Ionics 28-30 (1988) 1172-1175. https://doi.org/10.1016/0167-2738(88)90351-7

[43] J.B. Goodenough, Y. Kim, Challenges for rechargeable Li batteries, Chem. Mater. 22 (2010) 587-603. https://doi.org/10.1021/cm901452z

[44] H. Pan, Y.-S. Hu, L. Chen, Room-temperature stationary sodium-ion batteries for large-scale electric energy storage, EnergyEnviron. Sci. 6 (2013) 2338-2360. https://doi.org/10.1039/c3ee40847g

[45] A. Eftekhari, Potassium secondary cell based on Prussian blue cathode, J. Power Sour. 126 (2004) 221-228. https://doi.org/10.1016/j.jpowsour.2003.08.007

[46] D.P. DiVincenzo, E.J. Mele, Cohesion and structure in stage-1 graphite intercalation compounds, Phys.. Rev. B 32 (1985) 2538-2553. https://doi.org/10.1103/physrevb.32.2538

[47] D. Er, J. Li, M. Naguib, Y. Gogotsi, V.B. Shenoy, Ti_3C_2 MXene as a high capacity electrode material for metal (Li, Na, K, Ca) ion batteries, ACS Appl.Mater.Interfaces 6 (2014) 11173-11179. https://doi.org/10.1021/am501144q

[48] O. Mashtalir, M. Naguib, V.N. Mochalin, Y. Dall'Agnese, M. Heon, M.W. Barsoum, Y. Gogotsi, Intercalation and delamination of layered carbides and carbonitrides, Nat.Commun. 4 (2013) 1716. https://doi.org/10.1038/ncomms2664

[49] G. Mills, H. Jónsson, G.K. Schenter, Reversible work transition state theory: application to dissociative adsorption of hydrogen, Surf. Sci. 324 (1995) 305-337. https://doi.org/10.1016/0039-6028(94)00731-4

[50] Q. Tang, Z. Zhou, P. Shen, Are MXenes promising anode materials for Li-ion batteries? Computational studies on electronic properties and Li storage capability of Ti_3C_2 and $Ti_3C_2X_2$ (X= F, OH) monolayer, J. Am. Chem. Soc. 134 (2012) 16909-16916. https://doi.org/10.1021/ja308463r

[51] M.Q. Zhao, X. Xie, C.E. Ren, T. Makaryan, B. Anasori, G. Wang, Y. Gogotsi, Hollow MXene spheres and 3D macroporous MXene frameworks for Na ion storage, Adv.Mater. 29 (2017) 1702410. https://doi.org/10.1002/adma.201702410

[52] Y. Xie, M. Naguib, V.N. Mochalin, M.W. Barsoum, Y. Gogotsi, X. Yu, K.-W. Nam, X.-Q. Yang, A.I. Kolesnikov, P.R. Kent, Role of surface structure on Li-ion energy storage capacity of two-dimensional transition-metal carbides, J. Am. Chem. Soc. 136 (2014) 6385-6394. https://doi.org/10.1021/ja501520b

[53] Y. Xie, Y. Dall'Agnese, M. Naguib, Y. Gogotsi, M.W. Barsoum, H.L. Zhuang, P.R. Kent, Prediction and characterization of MXene nanosheet anodes for non-lithium-ion batteries, ACS Nano 8 (2014) 9606-9615. https://doi.org/10.1021/nn503921j

[54] J. Come, M. Naguib, P. Rozier, M.W. Barsoum, Y. Gogotsi, P.-L. Taberna, M. Morcrette, P. Simon, A non-aqueous asymmetric cell with a Ti2C-based two-dimensional negative electrode, J. Electrochem. Soc. 159 (2012) A1368-A1373. https://doi.org/10.1149/2.003208jes

[55] Q. Peng, J. Guo, Q. Zhang, J. Xiang, B. Liu, A. Zhou, R. Liu, Y. Tian, Unique lead adsorption behavior of activated hydroxyl group in two-dimensional titanium carbide, J. Am. Chem. Soc. 136 (2014) 4113-4116. https://doi.org/10.1021/ja500506k

MXenes: Fundamentals and Applications
Materials Research Foundations **51** (2019) 189-203

Materials Research Forum LLC
https://doi.org/10.21741/9781644900253-8

Chapter 8

MXenes for Biomedical Applications

Arka Bagchi[1], Solanki Sarkar[1], Ipsita Hazra Chowdhury[2], Arunima Biswas[1*]
and Sk Manirul Islam[2*]

[1] Cell and Molecular Biology Laboratory, Department of Zoology, University of Kalyani, Kalyani, Nadia, INDIA

[2] Department of Chemistry, University of Kalyani, Kalyani, Nadia, INDIA

arunima10@klyuniv.ac.in, arunima10@gmail.com, manir65@rediffmail.com

Abstract

MXenes have become an important family of 2-D layered materials having immense potential to be utilized in the fields of analytical chemistry, as a target for environmental monitoring and bio-medical applications. This chapter deals in detail with the diverse and far-reaching contribution of MXenes in biomedical sciences. Finally, a discussion is made on how these significant contributions of MXenes might have a long-term impact on biosensing and understanding disease biology. Lastly, the present lacuna in MXene research is also addressed and how future researches in this field might bring in high detection sensitivities.

Keywords

MXenes, Nanosheets, Biosensor, Imaging, Cytotoxicity, Therapeutics

Contents

1.   Introduction

MXenes, one of the youngest members of nano research, are a type of two-dimensional material, which is composed of metal carbides, nitrides or carbonitrides, sharing a common chemical formula of $M_{n+1}X_n$, where M refers to an early transition metal such as Ti, Zr, V, Ta, Nb, Mo, Cr and X is a carbon or nitrogen [1]. MXenes generally contain enough surface functional groups like oxygen, hydroxyl or fluorine along with the complete metal atomic layers. MXenes are a unique blend of metallic conductivity of some transition metal carbides/ nitrides and the hydrophilicity of some radicals and elements in it. Consequently, MXenes are provided with unique and useful electronic, optical and magnetic properties. Moreover, it is possible to theoretically predict the properties of MXenes by computational analyses due to the fact that they possess a highly ordered structure. The unique structure of MXenes provides them with unique properties compared to other two-dimensional materials, like graphene. Recent research over the past few years has seen path breaking discoveries responsible for the synthesis of novel MXenes which have higher chemical diversity and are structurally complex. Such complexity rarely existed in the other two-dimensional (2D) materials.

The structural and chemical complexities and diversities of MXenes provide an excellent field for fundamental and multidisciplinary research. Diverse and mostly environment friendly applications of MXenes have attracted attention of researchers in spite of the challenge of synthesizing MXenes with a uniform and pure surface termination.

One of the most unusual properties of MXenes compared to other popular two-dimensional materials is its electronic properties, which depend on the properties of the metal carbide with surface modulations. Scientists have demonstrated metallic to semiconductor-like properties in certain compositions of MXenes and few compositions like Ti_3C_2, Ti_2C, Mo_2C, etc. are shown to have electronic properties with mixed surface modifications [2,3]. It has been speculated that the outer metal layer and the method of preparation of MXenes significantly affect the electronic properties of MXenes, which may be attributed to their different level of defects and surface terminations [4]. If the defect is increased, the conductivity gradually decreases because of the destruction of the ordered structures, resulting in less free movement of the electrons. Moreover, the surface terminations of MXenes affect their ability to attract electrons and thereby affecting the conductivity of MXenes [5]. These electronic properties of MXenes are of great use in detecting several gases and reducing molecules with very high sensitivity. The semiconductor-like property of MXenes is also helpful for Reactive Oxygen Species generation and different biocatalysis reactions [6].

Researchers have demonstrated theoretically that MXenes interact with light and exhibit several unique optical properties such as light absorption, emission and scattering, which are very useful for their different applications in the biomedical field [7]. Magne and group have showed that the MXenes exhibits absorption in the UV-visible range [8]. There are also evidences that they have a wide range of absorption spectra from the UV-visible range to NIR region and this character of MXenes has been explored in the field of Photo Acoustic Imaging. It has also been shown to be useful in the field of Photo Thermal Therapy.

Another avenue of the biomedical applications of MXenes can be attributed to the magnetic property of MXenes, but this area is still the least explored area. Some researchers have theoretically demonstrated the magnetic properties of MXenes [9] and have proposed their use in the biomedical field. It has been theoretically shown by scientists that the crystal tension of MXenes greatly affect their magnetic property [10], but the use of the magnetic properties of MXenes is still restricted to the hybrid materials composed of different magnetic nanomaterials and MXenes due to lack of magnetic MXenes.

All these interesting and unique characters of MXenes have made scientists more interested about the applications of these materials, especially in the biomedical field.

2.   MXenes as antibacterial agent

Comparative study of graphene oxide and MXenes revealed that both can potentially exhibit antibacterial activity, MXenes being more effective than Graphene Oxide [6,11]. Moreover, many of the other metal oxides and nano particle (NP) formulations (Ag, ZnO and TiO_2), also show significant antibacterial properties and they function by the generation of ROS and through direct contact with bacterial membrane and can penetrate into the bacterial cell, interact with sulphur of proteins and phosphorous of DNA resulting in bacterial cell death[12–16]. There have been three forms of MXenes such as Ti_3AlC_2 (MAX), ML-MXene, delaminated $Ti_3C_2T_x$ nanosheets are used against different bacterial strains. Among them delaminated $Ti_3C_2T_x$ MXene have the highest growth inhibition for some bacterial strains [6,17]. MXenes affect the bacterial cell through membrane damage, cytoplasmic leakage as well as decrease the intra cellular densities. Due the structural stability of $Ti_3C_2T_x$ MXenes, it can directly enter the cell and release lactate dehydrogenase [6], which can be measured by the lactate dehydrogenase release assay, the release indicates the cellular damage of the bacterial cell. Moreover, MXenes show their antibacterial properties in suspension and as a membrane, and is extendable to other forms of MXenes.

Electrospun $Ti_3C_2T_z$ MXenes is used as a potent antibacterial agent in association with Chitosan(CS) having numerous biological applications such as antimicrobial activity, anti-inflammatory activity, biocompatibility, biodegradability, mucoadhesivity and many more [18,19]. Chitosan can act as antimicrobial agent by two types of mechanisms. The positively charged chitosan interacts with the negatively charged surface groups of cells that in turn affect the permeability of the cell, leading to cell death. Moreover, microbial RNA synthesis is also hampered by the chitosan [20]. When bacterial cells were treated with glutaraldehyde cross-linked with $Ti_3C_2T_Z$/CS composite nanofiber, it was observed that there was a reduction of *E.coli* as compared to normal control Chitosan fibers. On the other hand, the reduction was lower than in *S.aureus*. Since *S.aureus* is a gram-positive bacteria with a peptidoglycan cell wall with much greater width, this might be responsible for providing protection against $Ti_3C_2T_Z$/CS composite nanofiber antimicrobial activity. When $Ti_3C_2T_Z$/CS composite was administered to the bacterial cell colony, it was found through SEM micrograph that the morphological integrity has lost contact with $Ti_3C_2T_Z$/CS [18].

MXenes can also be used as antibacterial agent for wastewater management using titanium carbide ($Ti_3C_2T_x$). Polyvinylidene fluoride (PVDF) acts as a support for MXenes and such MXenes were utilized for wastewater and also water treatment [21,22]. A significant increase of antibacterial activity was observed in surface oxidized aged membrane as compared to freshly prepared membranes. *E.coli* and *B.subtilis* when exposed to $Ti_3C_2T_x$ modified membranes; the antibacterial activity of $Ti_3C_2T_x$ MXene was greater for *B.subtilis* and *E. coli* compared to control. But aged $Ti_3C_2T_x$ membranes exhibited more than 99% growth inhibitory effect on both bacteria under the same experimental conditions. $Ti_3C_2T_x$ MXene seemed to be more effective against gram-positive as well as gram-negative bacteria than micrometer-thick titanium carbide ($Ti_3C_2T_x$). Researchers have shown graphene oxide to be only 90% effective against *B.subtilis* [6]. MXenes have significant antibacterial properties, which were also substantiated by images obtained from scanning electron microscopy and atomic force microscopy. They showed potent actions against pathogenic waterborne bacteria at experimental levels. Hence, they seem to have immense potential to be used in water and wastewater treatment. Among all of the MXenes types, it has been proven that $Ti_3C_2T_Z$/CS nanofiber are most effective and also nontoxic element, whereas, at higher dose graphene oxide shows cytotoxic effect on mitochondrion and cell nuclei thus leading to apoptosis. Another nano-compound TiO_2 have a toxic effect on such major organs like brain and liver, causing acute necrosis, can also cross placental barrier harming the fetus [21].

3.   MXenes as biosensors

MXenes can detect very low concentrations of gases that were previously undetectable, thus it accounts for one of the most sensitive gas sensor. MXenes have unique property to detect indicators of ulcers and diabetes like ammonia and acetone even in trace amount compared to the presently available diagnostic biosensors. Since MXenes have a unique porous nature and chemical configuration, it is good for gaseous exchange, movement, or adsorbing. 2D MXene (Ti_2CO_2) monolayer have a high sensitivity and also have selectivity characteristic which make it a promising sensor for various gases like NH_3, H_2, CO, CO_2, NO_2, CH_4, N_2 [23]. TiC_2 is the thinnest phase among all MXenes groups. In cooling system, NH_3 is used as a substitute for chlorofluorocarbons that have toxic effect on the human body. Thus, it is necessary to find out and capture NH_3 gases for environmental controls [24]. Adsorption energy is low on Ti_2CO_2 of NH_3, which is suitable for adsorption as well as desorption of gases on a solid surface, thus Ti_2CO_2 can easily recovered after detecting NH_3. Introduction of two extra electrons to $ZrCO_2$ resulted in the release of NH_3 from MXene film and conversion of the process of chemisorption to physiorption, suggesting a promising role of MXene for gas sensing [23].

It is important in our vital life process to detect and quantify the most important monosaccharide, Glucose. Electrochemical transducer, signal-processing unit, recognition element for recording and amplification, are the important element for the electrochemical biosensors. Among them recognition element is the primary component of all biosensor to respond with the particular analyte within a large amount of other substances. According to the recognition elements, biosensors are classified as enzymatic, non-enzymatic and as immune sensors [25]. Among them enzymatic biosensor is highly selective and fast and also very much sensitive towards its molecule of interest [26]. Glucose biosensor is the most important and common biosensor. Glucose oxidase is the enzyme that is used to detect the glucose molecule [25–27]. The oxidation of glucose into gluconolatone and H_2O_2 is catalyzed by this enzyme. An efficient glucose biosensor could work with the electrochemical transducer. One of the most popular biosensor immobilization matrix in this field is glassy carbon electrode (GCE) [28]. There is a problem to use the glassy carbon electrode to detect glucose molecule, because of the oxidation/reduction reaction of H_2O_2. Many researchers have suggested using Au/nano carbon composite for glucose biosensing, where it can be utilized as electrochemical transducer. It has been also suggest the use Nafion solution for maximum adhesion of the enzyme molecule on the GCE. So there is Nafion solubilized Au/MXene/glucose oxidase over GCE have been fabricated. Reports suggest that Nafion solubilized Au/MXenes have super conductivity and catalytic activity than only the MXene/Nafion/glucose oxidase [26]. The mechanism behind the sensing of glucose molecule is that, glucose oxidase is made of one flavin adenine dinucleotide (FAD) and two identical protein subunit. The glucose oxidase enzyme houses the FAD molecule in its active site, where this FAD serves the purpose of a cofactor due to its excellent reversible electrochemistry. FAD is reduced and form $FADH_2$. Subsequently $FADH_2$ oxidized by dissolved O_2 and produce H_2O_2 and become FAD [29]. The following process represent by that equations:

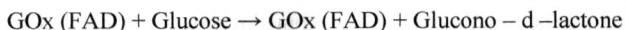

$$GOx\ (FAD) + Glucose \rightarrow GOx\ (FAD) + Glucono - d -lactone$$

$$GOx\ (FADH) + O_2 \rightarrow GOx\ (FAD) = H_2O_2$$

Combination of the following two equations which represent the oxidation process of glucose by Glucose oxidase enzyme is given below:

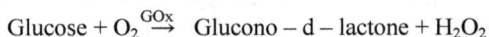

$$Glucose + O_2 \xrightarrow{GOx} Glucono - d - lactone + H_2O_2$$

It is widely reported that MXene nanosheets have excellent biosensing property when it is anchored by Au nanoparticals. Rakhi and her group [26] fabricated and proved that Nafion coating on the GOx/Au/ MXene/Nafion/GCE surface alleviate the interfering signals and raised the sensor functionality. These Nafion solubilized MXene compound can detect a very lower range of glucose molecule and are more effective than nanosheets without Au.

Zhou and group [30] have fabricated MXene (metal carbide) and chitosan based acetylcholinesterase biosensor for detecting organophosphate pesticides.

4. MXenes in bio-imaging

Mxenes have been proven to be good imaging contrast agents for bio-imaging, including photoacoustic imaging and luminescence imaging.

Photoacoustic imaging, an emerging diagnostic-imaging tool, can detect the induced pressure waves of laser-irradiated tissues, that is based on its low tissue attenuation coefficient, and thereby it is able to overcome the penetration limit of optical imaging. It also can detect real-time biological structures and functional information. For an efficient photoacoustic imaging, the contrast agent must have good photothermal conversion ability, so that it can show enhanced photoacoustic signal in contrast to the surrounding tissue. MXenes serve the function of a contrast agent very efficiently as they possess superior photothermal conversion ability [31]. The Localized Surface Plasmon Resonance effect of MXene nanosheets of semimetal character is the reason behind their strong absorption and conversion efficiency of broad spectra light. Scientists have demonstrated that various types of MXenes can convert the photon energy of excitation light into crystal vibrations and also release the energy macroscopically in form of thermal energy. Such photothermal effect has been observed in Ti_3C_2 MXenes [32]. Similar effects have also been observed in various other compositions of MXenes, such as Nb_2C and Ta_4C_3 [33,34]. Moreover, MXenes, as contrast agents, enhance the deep-tissue photoacoustic imaging, as they generally exhibit quite intensive absorption spectra.

Researchers have demonstrated that high photostability, desirable quantum yields, tunable wavelength and other efficient fluorescent properties for bio-imaging are evident in inorganic two dimensional nanomaterials and their corresponding quantum dots as compared to organic fluorescein. MXenes are also being used for luminescence cell imaging, courtesy to the recently developed MXene quantum dots. MXene quantum dots exhibit excitation-dependent luminescence, which is quite similar to the character of graphene and carbon dots. MXene flakes, when broken into very small dots by hydrothermal treatment, exhibits size-effect-induced quantum confinement and defect-induced luminescence [35]. Zhi and group [35] established that increase in the temperature of the hydrothermal treatment can significantly increase the luminescence of MXenes. It was speculated that the luminescence of the low temperature treated MXene dots arise mainly from the size confinement of the material, whereas increase in the temperature significantly increases the surface defects, thereby increasing the luminescence of MXene dots. Use of this luminescence property of MXene dots for cell imaging has been demonstrated by several groups of scientists in recent years. Researchers have also prepared MXene dots under a mild condition with the help of tetramethylammonium hydroxide (TMAOH), where there is no need of hydrothermal treatment but the dots are still able to exhibit photo stability and luminescence [36].

Recent researches in this area have also demonstrated the use of MXenes as X-ray/computed topography contrast agent beside their use as contrast agents in photoacoustic imaging and luminescence. For example, Ta_4C_3 MXenes have been used as contrast agents in X-ray/computed topography imaging [32]. MXene have also been used to monitor the neural activities by incorporating MXenes into a patterned scaffold for culturing neuron cells [37]. The neural activities can be monitored through the changes of the field effect of MXenes, as their conductivity is very sensitive to the surroundings.

Another new avenue of applications of MXenes refers to its use in Magnetic Resonance Imaging (MRI), which is an extensively applied tool for clinical diagnostic imaging due to the contrast difference of three dimensional tissues and also their high spatial

MXenes: Fundamentals and Applications   Materials Research Forum LLC
Materials Research Foundations 51 (2019) 189-203   https://doi.org/10.21741/9781644900253-8

resolution. One group of researchers has showed that Ti_3C_2 integration with MnO_x is beneficial to the Mn-based MRI [38].

5.   Therapeutic applications of MXenes

Recent researches on the biomedical applications of MXenes have revealed that these materials can be of great use as carriers to load cargo such as chemotherapeutic drugs for cancer treatment. Various surface modifications of MXenes enable them to load different types of cargos. For example, a very commonly used positively charged chemotherapeutic drug, doxorubicin, has been attached to the surface of negatively charged Ti_3C_2MXene to induce a synergistic therapy of doxorubicin-mediated chemotherapy and MXene mediated photothermal and photodynamic therapy [10]. MXenes have high surface area due to the two dimensional planar topology, which actually enables it to carry a versatile group of therapeutic agents on them, just like organic nanoplatforms [39] inorganic mesoporous carriers [40] and other 2D nanoplatforms [41,42]. Researchers have demonstrated that surface modification of MXenes directly affect the cargo carrying capacity of the MXenes. Scientists have also developed different methods where they utilized MXenes for pH responsive or on-demand drug release inside the cellular environment [43]. Scientists have developed Ti_3C_2-based nanoplatform (Ti_3C_2-DOX) which exhibits enormously high drug loading capacity. The cargo-carrying capacity of MXenes has also been utilized for carrying dyes to enhance bio-imaging.

Besides their use as cargo carriers, MXenes have also been used for photodynamic and photothermal killing of cells. Ti_3C_2 MXenes can act as pH sensors and are great drug loaders. Moreover, laser-triggered drug release capacity makes it an important component of targeted drug delivery system. Ti_3C_2 MXenes, have also been further explored for tumor eradication. One such example is the reactive oxygen species generation capability of Ti_3C_2 nanosheets [10]. It was speculated that the transfer of energy of photoexcited electrons from Ti_3C_2 to triplet oxygen is the driving force behind the production for reactive oxygen species. In vitro and in vivo studies with MXenes have showed generation of reactive oxygen species within the cellular environment where the unique electronic structure and photoelectronic properties of MXenes help them to serve as photosensitizers for photodynamic therapy. Scientists have doubted that the Localized Surface Plasmon Resonance effect may also play a pivotal role in reactive oxygen species generation. Though, the efficiency of MXenes for cell killing through the generation reactive oxygen species have been established, but the exact mechanism behind the scenario is still debatable.

MXenes: Fundamentals and Applications
Materials Research Foundations **51** (2019) 189-203

Materials Research Forum LLC
https://doi.org/10.21741/9781644900253-8

Moreover, not only for cargo delivery or targeted drug delivery, MXenes are presently experimentally used for breast cancer theranostics. Tantalum carbide (Ta_4C_3) MXene functionalized with ironoxide (Ta_4C_3-IONP-SPs composite MXenes) is reported by Yu Chen and his group in 2018 [44] is an superior nanoplatform for theranostics. Soybean phospholipid (SP) was used with Ta_4C_3-IONP MXenes to guarantee increased stability in biological mileu. Ta_4C_3-IONP-SPs exhibited good performance for contrast-enhanced CT imaging, which is mainly attributed to the Ta part of it and also the supermagnetic IONPs functioned as a remarkable contrast agent for T2-weighted MRI. These nanosheets could bring forth tumor regression in 4T1 tumor-bearing female BALB/c nude mice, verifying their highly efficient breast-tumor photo-ablation performance. MXene-based nanoplatforms (Ta4C3 MXene) are hence one of the most potent platform that can be exploited for theranostic approaches. Moreover, further functionalization strategies on MXenes might improve the platform to provide better biological applications. Such studies can be further explored for proper implementation of the same beyond experimental level. Researches have been conducted extensively with Fe_3O_4 NPs that are also emerging nanoplatforms for multimodality tumor imaging and therapy. The Fe_3O_4 NPs with active functional groups allow rational conjugation of biological and therapeutic molecules of interest. Recent and rapid developments in the synthesis and surface modification of Fe_3O_4 NPs also enable their use for effective theranostic applications [45]. But Fe_3O_4 NPs still remain a major challenge in the clinical utility because of the need for high NP concentrations. Studies with MXenes might be able to overcome this problem in the near future.

As discussed earlier that different types of MXenes possess high photothermal conversion efficiency and strong absorption in the NIR range, which is very much helpful for deep-tissue photothermal therapy. Researchers have demonstrated the use of different MXenes for *in vivo* photothermal therapy for tumors that opened up a new avenue of cancer treatment and research. It has been noted that scraped MXene dots are more efficient for photothermal therapy, which may be attributed to the more surface defects of MXene dots that can generate huge amount of heat [46].

Moreover, these MXenes are not only useful as therapeutic agents, but also have great value as a theranostic agent where they combine both diagnostic and therapeutic functions. Scientists are constantly trying to develop different ways to achieve better diagnostic and therapeutic capability with fewer side effects. The excellent photothermal conversion ability of MXenes and their simultaneous use as bioimaging contrast agents have provided researchers a new way to achieve their targets.

Discussion

Presently, MXenes, the youngest member of nano research is considerably popular due to their unique layered morphology and exotic metallic/catalytic properties. MXenes have excelled immensely as sensors having highly advanced detection schemes in multiple areas including health, environment, medicine etc. In spite of having various unique properties and high potential for biomedical applications, the extensive production of MXenes is still not achieved by scientists. There are several compositions of MXenes with unique characters, that are theoretically predicted, but scientists are yet to synthesize most of them.

MXenes, because of their large surface area and biocompatibility can be exploited to design advance nano(bio)hybrid systems with bioreceptors like aptamers (DNA/miRNA), antigen–antibody (Ag–Ab), whole cells, proteins and enzymes. These advancements will eventually help the researchers to develop more sophisticated and highly selective detection methodologies of various biomarkers that will be helpful for early detection of diseases and also for various diagnostic applications. Moreover, designing MXenes with low device dimension is a challenge that waits for the scientific committee for consideration. Fabrication of more advanced wearable strain sensors will be helpful to detect artery pulses and other ultra-weak physical stimuli.

Different approaches of cellular imaging techniques have also been facilitated by MXenes as they show good optical properties and moreover high biocompatibility. However, cytotoxicity of the same is a vital topic to be addressed before their commercial implementation. Further research is required for preparing MXenes with more control on the experimental conditions, so that it provides more control on the morphology, structure and surface terminations of the MXenes. Initial research on cancer cell lines has shown that MXenes have significant cytotoxicity towards cancerous cells. Moreover, increased reactive oxygen species (ROS) generation was observed with the application of MXenes on cancer cell line [47]. Though experimental results clearly indicate MXenes can be used for therapeutic approaches, extensive studies and further research are required to apply the same beyond the experimental level.

Further research is still required in developing more biocompatible and biodegradable MXenes. Several studies have already demonstrated that most of the MXenes, proposed for their biomedical applications, are biocompatible and some of them are biodegradable as well [48]. It is also evident from research that certain MXenes such as Ti_3C_2 can be classified as "practically nontoxic" based on the Acute Toxicity Rating Scale by Fish and Wildlife Service (FWS) [49]. To explore and exploit MXenes, researchers still have challenges to develop new compositions of MXenes and understand their applications in the biomedical field.

References

[1]   K. Huang, Z. Li, J. Lin, G. Han, P. Huang, Correction: Two-dimensional transition metal carbides and nitrides (MXenes) for biomedical applications, Chem. Soc. Rev. 47 (2018) 6889–6889. https://doi.org/10.1039/C8CS90090F

[2]   A. Lipatov, M. Alhabeb, M.R. Lukatskaya, A. Boson, Y. Gogotsi, A. sinitskii, effect of synthesis on quality, electronic properties and environmental stability of individual monolayer Ti_3C_2 mxene flakes, Advanced Electronic Materials. 2 (2016) 1600255. https://doi.org/10.1002/aelm.201600255

[3]   B. Anasori, C. Shi, E.J. Moon, Y. Xie, C.A. Voigt, P.R.C. Kent, S.J. May, S.J.L. Billinge, M.W. Barsoum, Y. Gogotsi, Control of electronic properties of 2D carbides (MXenes) by manipulating their transition metal layers, Nanoscale Horiz. 1 (2016) 227–234. https://doi.org/10.1039/C5NH00125K

[4]   L. Feng, X.-H. Zha, K. Luo, Q. Huang, J. He, Y. Liu, W. Deng, S. Du, Structures and mechanical and electronic properties of the Ti_2CO_2 MXene Incorporated with neighboring elements (Sc, V, B and N), Journal of Elec Materi. 46 (2017) 2460–2466. https://doi.org/10.1007/s11664-017-5311-5

[5]   X.-H. Zha, K. Luo, Q. Li, Q. Huang, J. He, X. Wen, S. Du, Role of the surface effect on the structural, electronic and mechanical properties of the carbide MXenes, EPL. 111 (2015) 26007. https://doi.org/10.1209/0295-5075/111/26007.

[6]   K. Rasool, M. Helal, A. Ali, C.E. Ren, Y. Gogotsi, K.A. Mahmoud, Antibacterial activity of Ti_3C_2Tx MXene, ACS Nano. 10 (2016) 3674–3684. https://doi.org/10.1021/acsnano.6b00181

[7]   G.R. Berdiyorov, Optical properties of functionalized $Ti_3C_2T_2$ (T = F, O, OH) MXene: First-principles calculations, AIP Advances. 6 (2016) 055105. https://doi.org/10.1063/1.4948799

[8]   D. Magne, V. Mauchamp, S. Célérier, P. Chartier, T. Cabioc'h, Spectroscopic evidence in the visible-ultraviolet energy range of surface functionalization sites in the multilayer ${\mathrm{Ti}}_{3}{\mathrm{C}}_{2}$ MXene, Phys. Rev. B. 91 (2015) 201409. https://doi.org/10.1103/PhysRevB.91.201409

[9]   A. Chandrasekaran, A. Mishra, A.K. Singh, Ferroelectricity, antiferroelectricity, and ultrathin 2D electron/hole gas in multifunctional monolayer MXene, Nano Lett. 17 (2017) 3290–3296. https://doi.org/10.1021/acs.nanolett.7b01035

[10]   G. Liu, J. Zou, Q. Tang, X. Yang, Y. Zhang, Q. Zhang, W. Huang, P. Chen, J. Shao, X. Dong, Surface modified Ti_3C_2 MXene nanosheets for tumor targeting photothermal/photodynamic/chemo synergistic therapy, ACS Appl Mater Interfaces. 9 (2017) 40077–40086. https://doi.org/10.1021/acsami.7b13421

[11]   S. Liu, T.H. Zeng, M. Hofmann, E. Burcombe, J. Wei, R. Jiang, J. Kong, Y. Chen, Antibacterial activity of graphite, graphite oxide, graphene oxide, and reduced graphene oxide: membrane and oxidative stress, ACS Nano. 5 (2011) 6971–6980. https://doi.org/10.1021/nn202451x

[12]   Y. Li, W. Zhang, J. Niu, Y. Chen, Mechanism of photogenerated reactive oxygen species and correlation with the antibacterial properties of engineered metal-oxide nanoparticles, ACS Nano. 6 (2012) 5164–5173. https://doi.org/10.1021/nn300934k

[13]   V. Lakshmi Prasanna, R. Vijayaraghavan, Insight into the mechanism of antibacterial activity of ZnO: surface defects mediated reactive oxygen species even in the dark, Langmuir. 31 (2015) 9155–9162. https://doi.org/10.1021/acs.langmuir.5b02266

[14]   Y.-W. Wang, A. Cao, Y. Jiang, X. Zhang, J.-H. Liu, Y. Liu, H. Wang, Superior antibacterial activity of zinc oxide/graphene oxide composites originating from high zinc concentration localized around bacteria, ACS Appl Mater Interfaces. 6 (2014) 2791–2798. https://doi.org/10.1021/am4053317

[15]   W. Zhang, Y. Li, J. Niu, Y. Chen, Photogeneration of reactive oxygen species on uncoated silver, gold, nickel, and silicon nanoparticles and their antibacterial effects, Langmuir. 29 (2013) 4647–4651. https://doi.org/10.1021/la400500t

[16]   O. Choi, Z. Hu, Size dependent and reactive oxygen species related nanosilver toxicity to nitrifying bacteria, Environ. Sci. Technol. 42 (2008) 4583–4588.

[17]   S. Chernousova, M. Epple, Silver as antibacterial agent: ion, nanoparticle, and metal, Angew. Chem. Int. Ed. Engl. 52 (2013) 1636–1653. https://doi.org/10.1002/anie.201205923

[18]   E.A. Mayerberger, R.M. Street, R.M. McDaniel, M.W. Barsoum, C.L. Schauer, Antibacterial properties of electrospun Ti_3C_2Tz (MXene)/chitosan nanofibers, RSC Adv. 8 (2018) 35386–35394. https://doi.org/10.1039/C8RA06274A

[19]   R. Niu, J. Qiao, H. Yu, J. Nie, D. Yang, Electrospun composite nanofibrous membrane as wound dressing with good adhesion, Front. Chem. China. 6 (2011) 221–226. https://doi.org/10.1007/s11458-011-0244-7

[20]   Y.-C. Chung, C.-Y. Chen, Antibacterial characteristics and activity of acid-soluble chitosan, Bioresour. Technol. 99 (2008) 2806–2814. https://doi.org/10.1016/j.biortech.2007.06.044

[21]   K. Rasool, K.A. Mahmoud, D.J. Johnson, M. Helal, G.R. Berdiyorov, Y. Gogotsi, Efficient antibacterial membrane based on two-dimensional Ti_3C_2Tx (MXene) Nanosheets, Scientific Reports. 7 (2017) 1598. https://doi.org/10.1038/s41598-017-01714-3

[22]   Y. Ying, Y. Liu, X. Wang, Y. Mao, W. Cao, P. Hu, X. Peng, Two-dimensional titanium carbide for efficiently reductive removal of highly toxic chromium(VI) from water, ACS Appl. Mater. Interfaces. 7 (2015) 1795–1803. https://doi.org/10.1021/am5074722

[23]   E. Lee, A. Vahid Mohammadi, B.C. Prorok, Y.S. Yoon, M. Beidaghi, D.-J. Kim, Room temperature gas sensing of two-dimensional titanium carbide (MXene), ACS Appl Mater Interfaces. 9 (2017) 37184–37190. https://doi.org/10.1021/acsami.7b11055

[24]   X. Yu, Y. Li, J. Cheng, Z. Liu, Q. Li, W. Li, X. Yang, B. Xiao, Monolayer Ti_3CO_2: A promising candidate for NH_3 sensor or capturer with high sensitivity and selectivity, ACS Appl Mater Interfaces. 7 (2015) 13707–13713. https://doi.org/10.1021/acsami.5b03737

[25]   A. Heller, B. Feldman, Electrochemical glucose sensors and their applications in diabetes management, Chemical Reviews. 108 (2008) 2482–2505. https://doi.org/10.1021/cr068069y

[26]   R.B. Rakhi, P. Nayak, C. Xia, H.N. Alshareef, Novel amperometric glucose biosensor based on MXene nanocomposite, Scientific Reports. 6 (2016) 36422. https://doi.org/10.1038/srep36422

[27]   C. Liu, Y. Sheng, Y. Sun, J. Feng, S. Wang, J. Zhang, J. Xu, D. Jiang, A glucose oxidase-coupled DNAzyme sensor for glucose detection in tears and saliva, BiosensBioelectron. 70 (2015) 455–461. https://doi.org/10.1016/j.bios.2015.03.070

[28]   M. Baghayeri, Glucose sensing by a glassy carbon electrode modified with glucose oxidase and a magnetic polymeric nanocomposite, RSC Adv. 5 (2015) 18267–18274. https://doi.org/10.1039/C4RA15888A

[29]   P. Nayak, P.N. Santhosh, S. Ramaprabhu, Synthesis of Au-MWCNT–graphene hybrid composite for the rapid detection of H_2O_2 and glucose, RSC Adv. 4 (2014) 41670–41677. https://doi.org/10.1039/C4RA05353B

[30]   L. Zhou, X. Zhang, L. Ma, J. Gao, Y. Jiang, Acetylcholinesterase/chitosan-transition metal carbides nanocomposites-based biosensor for the organophosphate pesticides detection, Biochemical Engineering Journal. 128 (2017) 243–249. https://doi.org/10.1016/j.bej.2017.10.008

[31]   H. Lin, X. Wang, L. Yu, Y. Chen, J. Shi, Two-dimensional ultrathin MXene ceramic nanosheets for photothermal conversion, Nano Lett. 17 (2017) 384–391. https://doi.org/10.1021/acs.nanolett.6b04339

[32]   H. Lin, S. Gao, C. Dai, Y. Chen, J. Shi, A two-dimensional biodegradable niobium carbide (MXene) for photothermal tumor eradication in NIR-I and NIR-II biowindows, J. Am. Chem. Soc. 139 (2017) 16235–16247. https://doi.org/10.1021/jacs.7b07818

[33]   C. Dai, H. Lin, G. Xu, Z. Liu, R. Wu, Y. Chen, Biocompatible 2D Titanium carbide (Mxenes) composite nanosheets for pH-responsive MRI-guided tumor hyperthermia, Chem. Mater. 29 (2017) 8637–8652. https://doi.org/10.1021/acs.chemmater.7b02441

[34]   C. Dai, Y. Chen, X. Jing, L. Xiang, D. Yang, H. Lin, Z. Liu, X. Han, R. Wu, Two-dimensional tantalum carbide (MXenes) composite nanosheets for multiple imaging-guided photothermal tumor ablation, ACS Nano. 11 (2017) 12696–12712. https://doi.org/10.1021/acsnano.7b07241

[35]   Q. Xue, H. Zhang, M. Zhu, Z. Pei, H. Li, Z. Wang, Y. Huang, Y. Huang, Q. Deng, J. Zhou, S. Du, Q. Huang, C. Zhi, Photoluminescent Ti_3C_2 MXene quantum dots for multicolor cellular imaging, Adv. Mater. Weinheim. 29 (2017). https://doi.org/10.1002/adma.201604847

[36]   Z. Wang, J. Xuan, Z. Zhao, Q. Li, F. Geng, Versatile cutting method for producing fluorescent ultrasmall Mxene sheets, ACS Nano. 11 (2017) 11559–11565. https://doi.org/10.1021/acsnano.7b06476

[37]   B. Xu, M. Zhu, W. Zhang, X. Zhen, Z. Pei, Q. Xue, C. Zhi, P. Shi, Ultrathin MXene-micropattern-based field-effect transistor for probing neural activity, Adv. Mater. Weinheim. 28 (2016) 3333–3339. https://doi.org/10.1002/adma.201504657

[38]   D. Ni, W. Bu, E.B. Ehlerding, W. Cai, J. Shi, Engineering of inorganic nanoparticles as magnetic resonance imaging contrast agents, ChemSoc Rev. 46 (2017) 7438–7468. https://doi.org/10.1039/c7cs00316a

[39]   L. Mei, Z. Zhang, L. Zhao, L. Huang, X.L. Yang, J. Tang, S.S. Feng, Pharmaceutical nanotechnology for oral delivery of anticancer drugs., Adv Drug Deliv Rev. 65 (2013) 880–890. https://doi.org/10.1016/j.addr.2012.11.005

[40]   Y. Chen, H. Chen, J. Shi, In vivo bio-safety evaluations and diagnostic/therapeutic applications of chemically designed mesoporous silica nanoparticles, Adv. Mater. Weinheim. 25 (2013) 3144–3176. https://doi.org/10.1002/adma.201205292

[41]   W. Yin, L. Yan, J. Yu, G. Tian, L. Zhou, X. Zheng, X. Zhang, Y. Yong, J. Li, Z. Gu, Y. Zhao, High-throughput synthesis of single-layer MoS_2 nanosheets as a near-

infrared photothermal-triggered drug delivery for effective cancer therapy, ACS Nano. 8 (2014) 6922–6933. https://doi.org/10.1021/nn501647j

[42]   Q. Weng, B. Wang, X. Wang, N. Hanagata, X. Li, D. Liu, X. Wang, X. Jiang, Y. Bando, D. Golberg, Highly water-soluble, porous, and biocompatible boron nitrides for anticancer drug delivery, ACS Nano. 8 (2014) 6123–6130. https://doi.org/10.1021/nn5014808

[43]   X. Han, J. Huang, H. Lin, Z. Wang, P. Li, Y. Chen, 2D Ultrathin MXene-based drug-delivery nanoplatform for synergistic photothermal ablation and chemotherapy of cancer, AdvHealthc Mater. 7 (2018) e1701394. https://doi.org/10.1002/adhm.201701394

[44]   Z. Liu, H. Lin, M. Zhao, C. Dai, S. Zhang, W. Peng, Y. Chen, 2D Superparamagnetic tantalum carbide composite mxenes for efficient breast-cancer theranostics, Theranostics. 8 (2018) 1648–1664. https://doi.org/10.7150/thno.23369

[45]   Y. Hu, S. Mignani, J.-P. Majoral, M. Shen, X. Shi, Construction of iron oxide nanoparticle-based hybrid platforms for tumor imaging and therapy, Chem. Soc. Rev. 47 (2018) 1874–1900. https://doi.org/10.1039/C7CS00657H

[46]   X. Yu, X. Cai, H. Cui, S.-W. Lee, X.-F. Yu, B. Liu, Fluorine-free preparation of titanium carbide MXene quantum dots with high near-infrared photothermal performances for cancer therapy, Nanoscale. 9 (2017) 17859–17864. https://doi.org/10.1039/c7nr05997c

[47]   L. Zong, H. Wu, H. Lin, Y. Chen, A polyoxometalate-functionalized two-dimensional titanium carbide composite MXene for effective cancer theranostics, Nano Res. 11 (2018) 4149–4168. https://doi.org/10.1007/s12274-018-2002-3

[48]   H. Lin, Y. Chen, J. Shi, Insights into 2D MXenes for versatile biomedical applications: current advances and challenges ahead, AdvSci (Weinh). 5 (2018). https://doi.org/10.1002/advs.201800518

[49]   G.K. Nasrallah, M. Al-Asmakh, K. Rasool, K.A. Mahmoud, Ecotoxicological assessment of Ti_3C_2Tx (MXene) using a zebrafish embryo model, Environ. Sci.: Nano. 5 (2018) 1002–1011. https://doi.org/10.1039/C7EN01239J

MXenes: Fundamentals and Applications

Materials Research Foundations **51** (2019) 204-215

Materials Research Forum LLC

https://doi.org/10.21741/9781644900253-9

Chapter 9

MXene and its Sensing Applications

Pramod K. Kalambate [a], Santosh W. Zote [b], Yue Shen [a], Dinesh N. Navale [c],
Dnyaneshwar K. Kulal [d], Jingyi Wu[a], Prasanna B. Ranade [c], Ramyakrishna Pothu [e],
Rajender Boddula [f], Yunhui Huang [a*]

[a] State Key Laboratory of Materials Processing and Die & Mould Technology, School of
Materials Science and Engineering, Huazhong University of Science and Technology, Wuhan,
Hubei 430074, PR China

[b] Department of Chemistry, PTVA's Sathaye College, Vile Parle (East), Mumbai 400 057, India

[c] Department of Chemistry, VES College, Chembur, Mumbai 400 071, India

[d] Department of Dyestuff Technology, Institute of Chemical Technology, Matunga, Mumbai 400
019, India

[e] College of Chemistry and Chemical Engineering, Hunan University, Changsha 410082, PR
China

[f] National Centre for Nanoscience and Technology, Beijing, 100190, PR China

* huangyh@hust.edu.cn

Abstract

MXenes possess excellent electrical conductivities, unique layered structure, large
surface area, and good catalytic activities which make them appropriate for the
fabrication of sensing platforms. This chapter discusses recent advances in the area of
MXene based electrochemical (bio) sensors, piezoresistive sensors, gas sensors, and
optical sensors. Each sensor has been discussed with suitable example in order to give a
clear idea about the utility of these materials in sensing. These sensors are useful for the
determination of a plethora of analytes, gases, and monitoring of the human subtle body
movements.

Keywords

MXene, Electrochemical Sensor, Piezoresistive Sensors, Electrocatalyst, Gas Sensors

Contents

1.   Introduction

MXenes are 2D layered carbides or carbonitrides obtained by selective etching of element Al from the parent MAX-Ti_3AlC_2. Recently, MXene has been used in the wide range of applications ranging from energy storage, sensing, to environmental remediation [1-3]. In spite of its small journey since its discovery, it has shown promising results in the wide areas of science and industries. Owing to its excellent electrical conductivities, unique layered structure, high specific surface area, ability to incorporate functional group and good biocompatibility, these materials are gaining huge interest in the sensing field. MXenes have shown great potential in several kinds of sensors viz., electrochemical, optical, gas sensors, and pressure sensors. A powerful feature of MXene is its ability to incorporate other nanomaterials. Hence, the addition of noble metal nanoparticles and other nanomaterials such as carbon nanotubes further improves the electrochemical properties of MXenes. As known, ultrathin 2D nanomaterials (graphene nanomaterials) have been largely studied for sensing purpose [4, 5]. However, there are some demerits associated with these materials, including low long-time stability and stacking due to ultrathin nature. On the other hand, chosen for its outstanding properties, MXenes are multi-layer structures exhibiting excellent long-time stability. We found that the MXene based electrochemical, optical and gas sensors have been utilized for the detection of a broad variety of analytes and gases with good sensitivity and selectivity. Additionally, MXene based piezoresistive sensors becoming popular for the monitoring of human subtle body movements. This chapter considers the current advances in the employment of MXenes and MXene-based nanocomposite in sensing devices. The underlying goal of this chapter is to give a clear idea about the recent advancement in MXene based sensing devices.

MXenes: Fundamentals and Applications   Materials Research Forum LLC
Materials Research Foundations **51** (2019) 204-215   https://doi.org/10.21741/9781644900253-9

2.   MXenes based sensors

MXenes exhibit outstanding chemical, physical, and electrical properties which are suitable for fabricating promising sensing platforms. So far, two MXene have been mostly utilized in electrochemical (bio), gas and optical sensing. This section focuses on the utility of MXenes for different kinds of sensors.

2.1   MXene for electrochemical (bio) sensing

Electrochemical sensors are popular detection tool and have been used for a long time for the determination of the range of analytes such as pharmaceutical drugs, toxic metals, pesticides, and so on [6-9]. The carbon nanomaterials, metal nanoparticle, metal oxides, and conducting polymers have been extensively studied for the fabrication of sensitive and selective electrochemical (bio) sensors [4, 10, 11]. Recently, MXenes have showcased big promise as a promising electrode modifier for the detection of several analytes. For example, Zhang et al., reported $Ti_3C_2T_x$-MXenes modified SPE for the simultaneous electrochemical determination of acetaminophen and isoniazid [12]. These two drugs are extensively used in the medicinal field for the treatment of diseases. However, excessive levels of these analytes in the body cause severe side effects. Monitoring the concentration of these drugs in body fluids provides valuable information for the optimization of treatment. Hence, the electrochemical sensor was developed for the simultaneous determination of these drugs in pharmaceutical and biological samples. In a typical study, the $Ti_3C_2T_x$ MXene was prepared from the parent MAX phase Ti_3AlC_2 using lithium fluoride in acidic conditions. MXene exhibits typical layered morphology (Fig.1a), and EDS spectra show corresponding elements in $Ti_3C_2T_x$ (Fig. 1b). Then, obtained MXene was drop casted on the screen-printed electrode and used for electrochemical measurements. CV study in $[FeCN_6]^{3-4-}$ redox couple showed that the peak current was enhanced and the potential difference of oxidation and reduction peak of MXene/SPE electrode was reduced indicating its suitability in electrochemical sensing (Fig. 1c). Electrochemical impedance spectroscopy of the MXene modified SPE showed charge transfer resistance (R_{ct}) of 132 Ω of [Fe CN6] 3-4- at MXene modified. 745 Ω for SPE. The diameter of the semicircle was decreased when SPE was modified with MXene showing excellent electron transfer property and better mass transfer performance (Fig. 1d). Similarly, the electrochemical response was significantly enhanced when acetaminophen and isoniazid were determined on the MXene/SPE electrode (Fig. 1e and d). DPV was used for individual and simultaneous determination of both analytes. The developed sensor exhibited a linear response towards oxidation of acetaminophen and isoniazid in the concentration range from 0.25-2000 μM and 0.1–4.6 mM with detection limits of 0.048 μM and 0.064 μM respectively. The sensor was used for the determination

Materials Research Forum LLC

https://doi.org/10.21741/9781644900253-9

of these drugs in pharmaceutical and blood serum samples with satisfactory recovery in the range from 99.1% to 102.5% indicating that no matrix effect from the blood serum samples interferes with the analysis. After two weeks the sensor maintained 97.2% and 95.8% of its initial response with %RSD of 2.7% and 3.5% for acetaminophen and isoniazid, respectively implying its good stability for long term use. Additionally, the reproducibility and repeatability of the MXene/SPE are good, and measurements can be carried out with good accuracy. Thus, such simple, sensitive, and selective sensor can be of great use for the detection of several analytes.

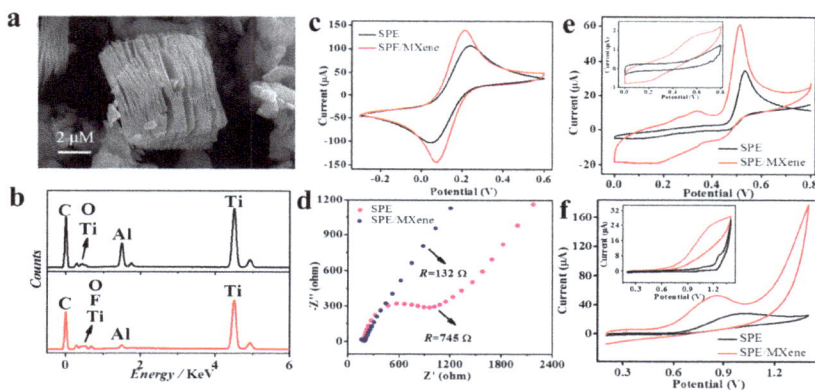

Figure 1. *a. SEM and b EDS of Ti_3AlC_2 before (black spectra) and after (red spectra) HF treatment; c CV and d EIS of SPE and SPE/MXene for 5 mM $[Fe(CN)_6]^{3-/4-}$ in 0.1 M KCl at 100 mV/s. EIS was performed in the frequency range from 0.1 Hz~100 kHz at an initialization potential of 0.224 V. CV of SPE and SPE/MXene for 1 mM e acetaminophen and f 1 mM isoniazid in 0.1 M H_2SO_4, respectively. Inset shows CV of SPE and SPE/MXene in the blank. [Reprinted with permission from ref. 12. Copyright 2019 Elsevier B.V].*

Similarly, based on the electrochemical approach, MXene or MXenes nanocomposite have been used for the analysis of several analytes. For example, Ti_3C_2–MXene has been used for tyrosinase immobilization and its use for the detection of phenol in real water samples [13]. Furthermore, $Ti_3C_2T_x$-MXene electrochemical microfluidic biosensor has been developed for the continuous assay of urea, uric acid, and creatinine [14]. In another study, $Ti_3C_2Tx/PtNPs/GCE$ was used for the sensing of H_2O_2 and small redox molecules such as ascorbic acid, uric acid, dopamine, and acetaminophen [2]. These reports

Materials Research Forum LLC
https://doi.org/10.21741/9781644900253-9

confirmed that the MXenes materials are promising electrode modifiers for the electrochemical sensors.

2.2   MXenes for optical sensing

Similar to electrochemical sensors, MXenes have been used for the development of optical sensors. For example, Desai et al. reported ultra small Ti_3C_2 nanosheets based fluorescence sensor, which could detect Ag^+ and Mn^+ ions with good sensitivity and selectivity [15]. The highest emission fluorescence peak was observed at 461nm at an excitation wavelength of 384 nm. The sensor showed a linear response for Ag^+ and Mn^+ ions in the range of 0.1–40μM and 0.5–60μM with LODs of 9.7 and 102 nM, respectively. The promising feature of this sensor is that the response was only quenched in the presence of these two metal ions. A strong interaction between metal ions and MXene surface is responsible for such selective response. The sensor could detect Ag^+ and Mn^+ in food and water samples with good recoveries from 99.8–101.9% and 98.54–103.95%, respectively. The %RSD for the simultaneous detection of Ag^+ and Mn^+ are 0.07-1.6% and 0.02-3.95%. Additionally, the interference study was conducted in the presence of commonly found metal ions and pesticides in which the response was unaffected, showing the promising application of this sensor. Thus, such low cost, accurate, and selective sensor opens up new ways for the trace level detection in environmental samples. In another example, Ti_3C_2 quantum dots have been used for the monitoring of intracellular pH values based on ratiometric photoluminescence [16]. The intracellular pH values are important and can be used as a signal for a variety of diseases and health of cells. The ratiometric pH sensor was developed by combining pH sensitive Ti_3C_2 quantum dots and pH insensitive $[Ru(dpp)_3]Cl_2$. Ti_3C_2 quantum dots revealed bright excitation-dependent blue photoluminescence response originating from the size effect and surface defects. The developed Ti_3C_2 quantum dots showed good biocompatibility and stability, high water dispersibility, and low toxicity, which allowed successful monitoring of intracellular pH values. In another study, Ti_3C_2-MXene has been used as off-on fluorescent nanoprobe for reliable detection of human papillomavirus [17].

2.3   MXene for gas sensing

Gas sensing devices have been employed for tracking/examining toxic gases in the environment [18]. Various kinds of gas sensors based on diverse sensing materials and methods have been investigated. The credential of gas sensors can be assessed by parameters such as selectivity, sensitivity, and response time, etc., [19, 20]. An ideal gas sensor must be highly sensitive, selective, stable and have low response time [21]. Since its discovery in 2011, MXenes have generated great interest in gas sensing as this material satisfies the requirement of an ideal sensor. $Ti_3C_2T_x$ was the first discovered and well-studied material of the MXenes. The sensing phenomenon of MXene is due to its metallic conductivity [22].

MXene gas sensing studies reveal that polar organic molecules those capable of weak hydrogen-bonding, give a small signal that too at high concentration limit. On the other hand, a non-polar molecule like toluene and cyclohexane, those without heteroatom give no response. Whereas, oxygen being non-polar inorganic molecules could not generate any response. Such high selectivity might be attributed to the strong electronic attraction between the target gas molecule and the MXene. The hydrophilic nature of MXenes play vital role in gas sensing. NH_3 is emitted by human as well as various industrial activities. Its detection at low concentration (down to ppb level) is a big task. NH_3 sensors are found use in various industries such as food technology, chemical plants, and medical diagnosis [23, 24]. The higher level of NH3 can be easily detected by its penetrating odour while low concentration requires sensing gas sensors. MXene materials can be regarded as solid-state gas sensors having low electrical noise and strong signal. A gas sensing performance of the $Ti_3C_2OH_2$-MXene was investigated at room temperature for several gases such as ammonia, acetone, ethanol, propanal, sulfur dioxide, and nitrogen oxide. The study revealed that ethanol had the highest response in comparision to other gases. The gas sensing response is also dependent on the thickness of the film [22].

$Ti_3C_2T_x$ showed p-type sensing behavior and could effectively detect acetone, ethanol, methanol, and ammonia gas at room temperature. Additionally, MXene sensor exhibits high regenerating ability towards detecting NH_3 by reversal capture and release of the gas [25]. Strong adsorption of polarized gases on the functionalized active sites/defects of $Ti_3C_2T_x$ caused the improvement of resistance within $Ti_3C_2T_x$ film. For NH_3, adsorption of the gas with terminal groups of $Ti_3C_2T_x$ such as O- or OH- resulted in the generation of electrons according to Eq. (1), [26] and -OH in Eq. (2).[27] Possible interaction of NH_3 with a terminal functional group on MXene surface has been illustrated below.

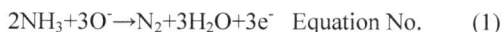

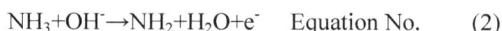

$$2NH_3 + 3O^- \rightarrow N_2 + 3H_2O + 3e^- \quad \text{Equation No.} \quad (1)$$

$$NH_3 + OH^- \rightarrow NH_2 + H_2O + e^- \quad \text{Equation No.} \quad (2)$$

NH_3 gets readily adsorbed on the defect sites of $Ti_3C_2T_x$ MXene than other gaseous molecules confirming MXene as an assuring interface for gas sensing applications. Representative MXenes $Ti_3C_2(OH)_2$ have excellent metallic conductivity [28], while the outer surface is completely covered with terminal functional groups, and surface functionalities are found to exist together [29]. Such a combination makes them an attractive material for gas sensing applications with a high SNR. The sensing ability of the developed device can be ascribed to the efficient adsorption/desorption of the gas molecules on the surface of $Ti_3C_2T_x$ 2D sheets. [30, 25] The target molecules are adsorbed on the surface of a sensing material by two types of adsorptive interactions. In the case of MXenes, gas adsorption takes place by interaction with terminal groups as well as at defects sites. Furthermore, the 3D-M based sensor was found to be sensitive to ultra-trace amount in the ppb range along with high sensitivity, LOD as low as 50 ppb, quick response, and recovery time.

2.4   MXene for piezoresistive sensing

Gao et al. reported a flexible piezoresistive sensor using Ti_3C_2-MXene exhibiting high sensitivity (Gauge Factor ~180.1), fast response of <30 ms, and extraordinary reversible compressibility [31]. Ti_3C_2-MXene was prepared by selective etching of element A from parent Ti_3AlC_2 using 50% HF. The SEM images showed a typical multilayer structure, and XRD results confirmed that the majority of the Ti_3AlC_2 had been converted to Ti_2C_2. After selective etching of Al with HF, the surface terminated on Ti_2C_2 consists of O and F functional groups. The piezoresistive sensor comprises of three layers in which the bottom layer is metal interdigital electrodes fabricated by ink printing and then magnetron sputtering metal on a flexible polyimide film (PI). Next, the PI substrate was coated with Ti_3C_2 layer by dipping and drying. Finally, the top layer of fabric with a hierarchical structure was coated on MXene layer as a protective layer. The whole device was encapsulated with polydiamethylsiloxane (PDMS) in order to improve stability. The practical application of the sensors was described by full range recognition of human activities. For example, the sensor was directly attached to an eye corner (Fig. 2a), the cheek (Fig. 2b) and the throat (Fig. 2c) to monitor the subtle motion resulting from microexpresssion. The change in current due to minimal strain change of eye blinking, cheek bulging, and throat swallowing are shown in Fig. 2. The relative I-T curves are different and are distinguishable with respect to shape and intensity change.

Additionally, the sensor was also used to monitor the bending-release movement of elbow, fingers, and ankles, as shown in Fig. 2d, e, and f. It is found that the current change of knee bending is the highest among various human motions and respective I-T curve is shown in Fig.2. The distance between two neighbouring interlayer's was greatly decreased under an external pressure, which causes a decrease in internal resistance (R_c) and an increase in conductivity. In situ TEM analysis showed that with the increase of external pressure the interlayer distance between neighbouring layers is decreased and when the pressure has removed the distance between both wider and narrow interlayer return back quickly suggesting good reversibility of MXene. However, the wider distance exhibit higher compress rate than narrow distance. These results proved that MXenes are promising materials for monitoring of human subtle body movements due to their good

Materials Research Forum LLC
https://doi.org/10.21741/9781644900253-9

flexibility and sensitivity. For a portable application, the sensor was coupled in series to a microcircuit implanted with a Bluetooth which transformed the change in current or resistance into a wireless electromagnetic signal (Fig. 2h). The knee motion from a Bluetooth connected system resembles the Agilent equipment both in terms of amplitude and shape but in a faster frequency than in Fig. 2g. In this study, the conductivity of MXene was found to 6, 500 S cm^{-1}, which is greater than graphene and CNTs. The sensor showed little signal attenuation after 4000 cycles confirming high stability of the developed sensor. Hence, such low cost, a highly flexible sensor with fast response, good stability, and an application for monitoring of subtle human activities makes a highly promising sensor. Similarly, 3D hybrid porous $Ti_3C_2T_x$-MXene sponge-network and 3D MXene/reduced graphene oxide aerogel based piezoresistive sensors were developed for monitoring of human physiological signals (respiration, joint movements, and pulses), and health activities [32, 33].

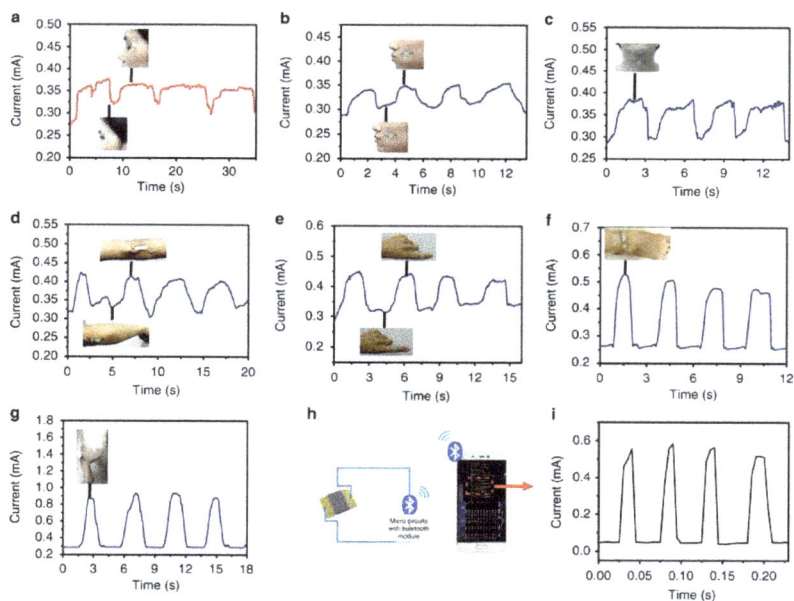

*Figure. 2. The current change of the developed sensor recorded to monitor eye blinking **a** cheek bulging **b** and throat swallowing **c**. The current change of the sensor recorded for the bending-release movement of the elbow **d**, fingers **e**, ankle **f**, and knee **g**. **h** the sensor connected to the micro circuit. **i** The response from **h** is quite similar to Agilent equipment in terms of amplitude and shape. Reprinted with permission from ref. 32. Copyright 2017 Nature Publishing Group].*

Conclusion

In summary, MXenes and their nanocomposites have shown great promise in different sensing devices, including electrochemical, optical, gas, and piezoresistive sensors. The unique layered morphology, excellent electrical conductivities, biocompatibilities, and hydrophilicity make them suitable for detection of the plethora of analytes and gases along with high sensitivity and selectivity. MXenes offer a high surface for functionalization with specific functional groups/enzymes without compromising properties which can be used for specific sensing applications. MXenes showed high reversibility and flexibility for fabricating piezoresistive sensors in the wide pressure ranges, which make them suitable for monitoring human physiological signals with great accuracy. So far, only a few MXenes (titanium carbide) have been utilized for sensing purpose, therefore, extensive research is required in order to see the utility of other MXenes in the sensing field.

Abbreviations

2D: two-dimensional; 3D: three-dimensional; SPE: screen printed electrode; LOD: lowest limit of detection; R_{ct}: charge transfer resistance; CNTs: carbon nanotubes; PtNPs: platinum nanoparticles; SEM: scanning electron microscopy; TEM: transmission electron microscopy; XRD: X-ray diffraction; SNR: signal to noise ratio

References

[1] S. Sun, Z. Xie, Y. Yan, S. Wu, Hybrid energy storage mechanisms for sulfur-decorated Ti_3C_2 MXene anode material for high-rate and long-life sodium-ion batteries, Chem. Eng. J. 366 (2019) 460-467. https://doi.org/10.1016/j.cej.2019.01.185

[2] L. Lorencova, T. Bertok, J. Filip, M. Jerigova, D. Velic, P. Kasak, K. A. Mahmoud, J. Tkac, Highly stable $Ti_3C_2T_x$ (MXene)/Pt nanoparticles-modified glassy carbon electrode for H_2O_2 and small molecules sensing applications, Sens. Actuators B Chem 263 (2018) 360-368. https://doi.org/10.1016/j.snb.2018.02.124

[3] Y. Zhang, L. Wang, N. Zhang, Z. Zhou, Adsorptive environmental applications of MXene nanomaterials: a review, RSC Adv. 8 (2018) 19895-19905. https://doi.org/10.1039/c8ra03077d

[4] P. K. Kalambate, B. J. Sanghavi, S. P. Karna, A. K. Srivastava, Simultaneous voltammetric determination of paracetamol and domperidone based on a graphene/platinum nanoparticles/nafion composite modified glassy carbon electrode,

Sens. Actuators B Chem 213 (2015) 285-294.
https://doi.org/10.1016/j.snb.2015.02.090

[5] P. K. Kalambate, C. R. Rawool, A. K. Srivastava, Voltammetric determination of pyrazinamide at graphene-zinc oxide nanocomposite modified carbon paste electrode employing differential pulse voltammetry, Sens. Actuators B Chem 237 (2016)196-205. https://doi.org/10.1016/j.snb.2016.06.019

[6] P. K. Kalambate, A. K. Srivastava, Simultaneous voltammetric determination of paracetamol, cetirizine and phenylephrine using a multiwalled carbon nanotube-platinum nanoparticles nanocomposite modified carbon paste electrode, Sens. Actuators B Chem 233 (2016) 237-248. https://doi.org/10.1016/j.snb.2016.04.063

[7] P. K. Kalambate, Y. Li, Y. Shen, Y. Huang, Mesoporous Pd@Pt core-shell nanoparticles supported on multi-walled carbon nanotubes as a sensing platform: Application to simultaneous electrochemical detection of anticancer drugs doxorubicin and dasatinib, Anal. Methods 11 (2019) 443-453. https://doi.org/10.1039/c8ay02381f

[8] B. J. Sanghavi, N. S. Gadhari, P. K. Kalambate, S. P. Karna, A. K. Srivastava, Potentiometric stripping analysis of arsenic using a graphene paste electrode modified with thiacrown ether and gold nanoparticles, Microchim. Acta 182 (2015) 1473-1481. https://doi.org/10.1007/s00604-015-1470-3

[9] X. Gao, Y. Gao, C. Bian, H. Ma, H. Liu, Electroactive nanoporous gold driven electrochemical sensor for the simultaneous detection of carbendazim and methyl parathion, Sens. Actuators B Chem 310 (2019) 78-85. https://doi.org/10.1016/j.electacta.2019.04.120

[10] P. K. Kalambate, Dhanjai, Z. Huang, Y. Li, Y. Shen, M. Xie, Y. Huang, A. K. Srivastava, Core@Shell nanomaterials based sensing devices: A Review, Trends Anal. Chem. 115 (2019) 147-161. https://doi.org/10.1016/j.trac.2019.04.002

[11] P. K. Kalambate, C. R. Rawool, A. K. Srivastava, Fabrication of graphene-multiwalled carbon nanotubes-polyaniline modified carbon paste electrode for simultaneous electrochemical determination of terbutaline sulphate and guaifenesin, New. J. Chem. 41 (2017) 7061-7072. https://doi.org/10.1039/c7nj00101k

[12] Y. Zhang, X. Jiang, J. Zhang, H, Zhang, Y. Li, Simultaneous voltammetric determination of acetaminophen and isoniazid using MXene modified screen-printed electrode, Biosens. Bioelectron. 130 (2019) 315-321. https://doi.org/10.1016/j.bios.2019.01.043

[13] L. Wu, X. Lu, Dhanjai, Z. –S. Wu, Y. Dong, X. Wang, S. Zheng, J. Chen, 2D transition metal carbide MXene as a robust biosensing platform for enzyme immobilization and ultrasensitive detection of phenol, Biosens. Bioelectron. 107 (2018) 69-75. https://doi.org/10.1016/j.bios.2018.02.021

[14] J. Liu, X. Jiang, R. Zhang, Y. Zhang, L. Wu, W. Lu, J. Li, Y. Li, H. Zhang, MXene-enabled electrochemical microfluidic biosensor: applications toward multicomponent continuous monitoring in whole blood, Adv. Funct. Mater. 29 (2019) 1807326. https://doi.org/10.1002/adfm.201807326

[15] M. L. Desai, H. Basu, R. K. Singhal, S. Saha , S. K. Kailasa, Ultra-small two dimensional MXene nanosheets for selective and sensitive fluorescence detection of Ag^+ and Mn^{2+} ions, Colloids Surf. A Physicochem. Eng. Asp. 565 (2019) 70–77. https://doi.org/10.1016/j.colsurfa.2018.12.051

[16] X. Chen, X. Sun, W. Xu, G. Pan, D. Zhou, J. Zhu, H. Wang, X. Bai, B. Dong, H. Song, Ratiometric photoluminescence sensing based on Ti_3C_2 MXene quantum dots as an intracellular pH sensor, Nanoscale 10 (2018) 1111–1118. https://doi.org/10.1039/c7nr06958h

[17] X. Peng, Y. Zhang, D. Lu, Y. Guo, S. Guo, Ultrathin Ti_3C_2 nanoscheets based off-on flurescent nanoprobe for rapid and sensitive detection of HPV infection, Sens. Actuators B Chem 286 (2019) 222-229. https://doi.org/10.1016/j.snb.2019.01.158

[18] T. Anukunprasert, C. Saiwan, E. Traversa, The development of gas sensor for carbon monoxide monitoring using nanostructure $Nb-TiO_2$, Sci. Technol. Adv. Mater. 6 (2005) 359–363. https://doi.org/10.1016/j.stam.2005.02.020

[19] G. Korotcenkov, Metal oxides for solid-state gas sensors: What determines our choice? Mater. Sci. Eng. B 139 (2007) 1–23.

[20] X. Liu, S. Cheng, H. Liu, S. Hu, D. Zhang, H. Ning, A survey on gas sensing technology, Sensor 12 (2012) 9635–9665. https://doi.org/10.3390/s120709635

[21] V. E. Bochenkov, G. B. Sergeev, Sensitivity, selectivity and stability of gas-sensitive metal oxide nanostructures in: A. Umar, Y. B. Hahn (Eds.), Metal oxide nanostructures and their applications, American Scientific Publication, 3 (2010), pp. 31–52.

[22] S. J. Kim, H.-J. Koh, C. E. Ren, O. Kwon, K. Maleski, S.-Y. Cho, B. Anasori, C.-K. Kim, Y.-K. Choi, J. Kim, Y. Gogotsi, H.-T. Jung, Metallic $Ti_3C_2T_x$ MXene gas sensors with ultrahigh signal-to-noise ratio, ACS Nano12 (2018) 986-993. https://doi.org/10.1021/acsnano.7b07460

[23] A. K. Prasad, D. J. Kubinskib, P. I. Gouma, Comparison of sol-gel and ion beam deposited MoO_3 thin film gas sensors for selective ammonia detection, Sens. Actuators B Chem 93 (2003) 25–30. https://doi.org/10.1016/s0925-4005(03)00336-8

[24] B. Timmer, W. Olthuis, A.V.D. Berg, Ammonia sensors and their application – a review, Sens. Actuators B Chem 107 (2005) 666–677. https://doi.org/10.1016/j.snb.2004.11.054

[25] B. Xiao, Y. Li, X. Yu, J. Cheng, MXenes: reusable materials for NH_3 sensor or capturer by controlling the charge injection, Sens. Actuators B Chem 235 (2016) 103-109. https://doi.org/10.1016/j.snb.2016.05.062

[26] M. Gautam, A. H. Jayatissa, Ammonia gas sensing behavior of graphene surface decorated with gold nanoparticles, Solid-State Electron. 78 (2012) 159-165. https://doi.org/10.1016/j.sse.2012.05.059

[27] L. Fenn, D. Kissel, Ammonia volatilization from surface applications of ammonium compounds on calcareous soils: I. general theory, Soil Sci. Soc. Am. J. 37 (1973) 855-859. https://doi.org/10.2136/sssaj1973.03615995003700060020x

[28] Y. Xie, P. R. C. Kent, Hybrid density functional study of structural and electronic properties of functionalized $Ti_n^+1X_n$ (X = C, N) monolayers, Phys. Rev. B: Condens, Matter Mater. Phys. 87 (2013) 235441. https://doi.org/10.1103/physrevb.87.235441

[29] F. Shahzad, M. Alhabeb, C. B. Hatter, B. Anasori, S. M. Hong, C. M. Koo, Y. Gogotsi, Electromagnetic interference shielding with 2D transition metal carbides (MXenes), Science 353 (2016) 1137−1140. https://doi.org/10.1126/science.aag2421

[30] X.-F. Yu, Y.-C. Li, J.-B. Cheng, Z.-B. Liu, Q.-Z. Li, W.-Z. Li, X. Yang, B. Xiao, Monolayer Ti_2CO_2: A promising candidate for NH_3 sensor or capturer with high sensitivity and selectivity, ACS Appl. Mater. Interfaces 7 (2015) 13707-13713. https://doi.org/10.1021/acsami.5b03737

[31] Y. Ma, N. Liu, L. Li, X. Hu, Z. Zou, J. Wang, S. Luo, Y. Gao, A highly flexible and sensitive piezoresistive sensor based on MXene with greatly changed interlayer distances, Nat. Commun. 8 (2017) 1207. https://doi.org/10.1038/s41467-017-01136-9

[32] Y. Yue, N. Liu, W. Liu, M. Li, Y. Ma, C. Luo, S. Wang, J. Rao, X. Hu, J. Su, Z. Zhang, Q. Huang, Y. Gao, 3D hybrid porous MXene-sponge network and its application in piezoresistive sensor, Nano Energy 50 (2018) 79–87. https://doi.org/10.1016/j.nanoen.2018.05.020

[33] Y. Ma, Y. Yue, H. Zhang, F.Cheng, W. Zhao, J. Rao, S. Luo, J.Wang, X. Jiang, Z. Liu, N. Liu, Y.Gao, 3D synergistical MXene/reduced graphene oxide aerogel for a piezoresistive sensor, ACS Nano 12 (2018) 3209−3216. https://doi.org/10.1021/acsnano.7b06909

Keyword Index

About the Editors

Dr. Inamuddin is currently working as Assistant Professor in the Chemistry Department, Faculty of Science, King Abdulaziz University, Jeddah, Saudi Arabia. He is a permanent faculty member (Assistant Professor) at the Department of Applied Chemistry, Aligarh Muslim University, Aligarh, India. He obtained Master of Science degree in Organic Chemistry from Chaudhary Charan Singh (CCS) University, Meerut, India, in 2002. He received his Master of Philosophy and Doctor of Philosophy degrees in Applied Chemistry from Aligarh Muslim University (AMU), India, in 2004 and 2007, respectively. He has extensive research experience in multidisciplinary fields of Analytical Chemistry, Materials Chemistry, and Electrochemistry and, more specifically, Renewable Energy and Environment. He has worked on different research projects as project fellow and senior research fellow funded by University Grants Commission (UGC), Government of India, and Council of Scientific and Industrial Research (CSIR), Government of India. He has received Fast Track Young Scientist Award from the Department of Science and Technology, India, to work in the area of bending actuators and artificial muscles. He has completed four major research projects sanctioned by University Grant Commission, Department of Science and Technology, Council of Scientific and Industrial Research, and Council of Science and Technology, India. He has published 138 research articles in international journals of repute and eighteen book chapters in knowledge-based book editions published by renowned international publishers. He has published forty-two edited books with Springer, United Kingdom, Elsevier, Nova Science Publishers, Inc. U.S.A., CRC Press Taylor & Francis Asia Pacific, Trans Tech Publications Ltd., Switzerland and Materials Research Forum LLC, U.S.A. He is the member of various editorial boards of the journals and serving as associate editor for journals such as Environmental Chemistry Letter, Applied Water Science, Euro-Mediterranean Journal for Environmental Integration, Springer-Nature, Frontiers Section Editor of Current Analytical Chemistry, published by Bentham Science Publishers, editorial board member for Scientific Reports-Nature and editor for Eurasian Journal of Analytical Chemistry. He has attended as well as chaired sessions in various international and national conferences. He has worked as a Postdoctoral Fellow, leading a research team at the Creative Research Initiative Center for Bio-Artificial Muscle, Hanyang University, South Korea, in the field of renewable energy, especially biofuel cells. He has also worked as a Postdoctoral Fellow at the Center of Research Excellence in Renewable Energy, King Fahd University of Petroleum and Minerals, Saudi Arabia, in the field of polymer electrolyte membrane fuel cells and computational fluid dynamics of polymer electrolyte membrane fuel cells. He is a life member of the Journal of the Indian

Chemical Society. His research interest includes ion exchange materials, a sensor for heavy metal ions, biofuel cells, supercapacitors and bending actuators.

Dr. Rajender Boddula is currently working as Chinese Academy of Sciences-President's International Fellowship Initiative (CAS-PIFI) at National Center for Nanoscience and Technology (NCNST, Beijing). His academic honors include University Grants Commission National Fellowship and many merit scholarships, study-abroad fellowships from Australian Endeavour Research fellowship and CAS-PIFI. He has published many scientific articles in international peer-reviewed journals and has authored six book chapters, and also serving as editorial board member and referee for reputed international peer-reviewed journals. His specialized areas of energy conversion and storage, which include nanomaterials, graphene, polymer composites, heterogeneous catalysis, photoelectrocatalytic water splitting, biofuel cell, and supercapacitors.

Prof. Abdullah M. Asiri is the Head of the Chemistry Department at King Abdulaziz University since October 2009 and he is the founder and the Director of the Center of Excellence for Advanced Materials Research (CEAMR) since 2010 till date. He is the Professor of Organic Photochemistry. He graduated from King Abdulaziz University (KAU) with B.Sc. in Chemistry in 1990 and a Ph.D. from University of Wales, College of Cardiff, U.K. in 1995. His research interest covers color chemistry, synthesis of novel photochromic and thermochromic systems, synthesis of novel coloring matters and dyeing of textiles, materials chemistry, nanochemistry and nanotechnology, polymers and plastics. Prof. Asiri is the principal supervisors of more than 20 M.Sc. and six Ph.D. theses. He is the main author of ten books of different chemistry disciplines. Prof. Asiri is the Editor-in-Chief of King Abdulaziz University Journal of Science. A major achievement of Prof. Asiri is the discovery of tribochromic compounds, a new class of compounds which change from slightly or colorless to deep colored when subjected to small pressure or when grind. This discovery was introduced to the scientific community as a new terminology published by IUPAC in 2000. This discovery was awarded a patent from European Patent office and from UK patent. Prof. Asiri involved in many committees at the KAU level and on the national level. He took a major role in the advanced materials committee working for KACST to identify the national plan for science and technology in 2007. Prof. Asiri played a major role in advancing the chemistry education and research in KAU. He has been awarded the best researchers from KAU for the past five years. He also awarded the Young Scientist Award from the Saudi Chemical Society in 2009 and also the first prize for the distinction in science from the Saudi Chemical Society in 2012. He also received a recognition certificate from the American Chemical Society (Gulf region Chapter) for the advancement of chemical science in the Kingdome. He received a Scopus certificate for the most publishing scientist in Saudi Arabia in chemistry in 2008. He is also a member of the editorial board of various journals of international repute. He is the Vice- President of Saudi Chemical Society (Western Province Branch). He holds four USA patents, more than one thousand publications in international journals, several book chapters and edited books.